Cartophilia

CARTOPHILIA

Maps and the Search for Identity
in the French-German Borderland

Catherine Tatiana Dunlop

The University of Chicago Press
Chicago and London

Catherine Tatiana Dunlop is assistant professor of modern European history at Montana State University, Bozeman.

The University of Chicago Press, Chicago 60637
The University of Chicago Press, Ltd., London
© 2015 by The University of Chicago
All rights reserved. Published 2015.
Printed in the United States of America

24 23 22 21 20 19 18 17 16 15   1 2 3 4 5

ISBN-13: 978-0-226-17302-3 (cloth)
ISBN-13: 978-0-226-17316-0 (e-book)
DOI: 10.7208/chicago/9780226173160.001.0001

Portions of chapter 3 first appeared in "Mapping a New Kind of European Boundary: The Language Border between Modern France and Germany," *Imago Mundi* 65 (2013). Reprinted with permission. Portions of chapter 4 first appeared in "Mapping Locality in a European Borderland: The Cartographic Construction of Identity, Space, and Boundaries in Alsace," in *Place and Locality in Modern France*, edited by Philip Whalen and Patrick Young (Bloomsbury Publishing, London, 2014). Reprinted with permission.

Library of Congress Cataloging-in-Publication Data
Dunlop, Catherine Tatiana, author.
  Cartophilia : maps and the search for identity in the French-German borderland / Catherine Tatiana Dunlop.
    pages : maps ; cm
  Includes bibliographical references and index.
  ISBN 978-0-226-17302-3 (cloth : alk. paper) —
ISBN 978-0-226-17316-0 (e-book)
  1. Alsace (France)—Maps—History—19th century.   2. Lorraine (France)—Maps—History—19th century.   3. Alsace (France)—Maps—History—20th century.   4. Lorraine (France)—Maps—History—20th century.   5. France—Boundaries—Germany—History.   6. Germany—Boundaries—France—History.   7. Cartography—Political aspects—France.   8. Cartography—Political aspects—Germany.   I. Title.
  GA865.A45D86 2015
  526.0944'3809034—dc23
                                                                        2014036322

For my parents,
John and Olga Dunlop

CONTENTS

AN AERIAL PERSPECTIVE

Every spring, the people of Alsace can look up at the sky and spot flocks of elegant, long-legged white storks returning home from their migration to the coasts of North Africa. Soaring above the landlocked region in the heart of Europe, the birds see the landscape below them as a burst of different colors: bright blue, golden yellow, and deep green. To the east, they see the fabled Rhine River, having recently emerged from Lake Constance, lazily snaking its way north toward Germany's Rhineland. On the left bank of the river, they catch sight of Strasbourg, the capital of Alsace, marked by its distinctive, single cathedral spire, which reaches piercingly into the sky. To the west of Strasbourg, the storks pass over the Alsatian Plain, dotted with picturesque wine-growing villages, each radiating around a tiny church tower. Villagers welcome the birds' arrival as a sign of good luck, erecting landing spots on the roofs of their half-timbered houses so the storks can build nests for their eggs. The birds that fly still further to the west approach the French interior, where the topography rises steeply in elevation, becoming the Vosges Mountains. An old, dying mountain chain, the Vosges crests resemble waves rolling across the horizon, obscured in parts by the clouds and fog trapped between them.

The storks that have migrated to Alsace every year, for the last several hundred years, would have been unaware that the region they have chosen as their seasonal home was the locus of a long-standing border conflict between two powerful human societies. An inland territory with abundant natural resources and a central location at the crossroads of states, languages, religions, and commercial networks, Alsace became a place for

Figure Intro.1 Map of Europe with current political borders. Drawn by Michele Mayor Angel.

Europeans of different backgrounds to exchange goods and ideas. But its strategically vital position between the Rhine and the Vosges, and French and German-speaking civilizations, led powerful leaders in Paris and Berlin to view Alsace, and parts of its neighboring region, Lorraine, as territorial objects to be conquered, possessed, and fiercely defended. Over the course of their history, Alsace and Lorraine's geographic location has been both a blessing and a burden for their people, existing alternately as a site of mediation and contestation between two of Europe's most powerful states.

While the conflict over the French-German borderland dates to Roman times, the modern territorial dispute first exploded with the guns of war in 1648. In that year, France's Sun King, Louis XIV, captured parts of Alsace from the German Holy Roman Empire, binding his new territory to the rest of the French kingdom with an iron curtain of star-shaped fortresses along the Rhine frontier. Lorraine entered into the French realm more peacefully, through diplomatic negotiations in 1766. Over a century later, in 1871, the newly unified German Empire, led by Emperor Wilhelm I, recaptured Alsace and part of Lorraine from the French, establishing a new border whose architecture consisted of fifty-five hundred stone markers that lined the crests of the Vosges. In the twentieth century, war determined the border's fate once again, this time by the overwhelming force of the two world wars. In 1918, 1940, and 1945, the French-German border—the most militarized in Europe—moved hundreds of kilometers back and forth between the shores of the Rhine and the mountaintops of the Vosges, as sovereignty over Alsace and Lorraine passed from German to French hands.

Figure Intro.2 Map of Alsace-Lorraine. Drawn by Michele Mayor Angel.

The shifting borders between modern France and Germany would not have been visible on the sweeping landscape that the region's migrating storks observe from hundreds of meters above the ground. From an aerial perspective, the territories that we now call France, Lorraine, Alsace, and Germany flow together into a continuous geological canvas. The man-made borders that have cut through the interlocking territories were conceived in the halls of European palaces, where French and German leaders met behind closed doors to decide the fate of a distant piece of land on the fringes of their realms. But over time, the French-German border came to represent much more than the sovereign limit between two increasingly powerful states; it transformed into a symbolic divide between two national cultures. In both France and Germany, citizens fixated on Alsace and Lorraine as coveted pieces of land without which their national territories felt incomplete and unfulfilled. But looking at the disputed terrain through the eyes of its migrating storks reminds us that seeing borders does not come naturally. It requires a socially constructed understanding of space. To train themselves to see the limits of their states, nations, regions, and "homelands" in the landscape of Alsace and Lorraine, Europeans needed a visual tool capable of combining a bird's sweeping view of land with the power of the human imagination. They needed maps.

AN AGE OF CARTOPHILIA

It is no coincidence that the age of European nationalism, which gave rise to the violent border conflict between modern France and Germany, was also the age of a passionate and widespread cartophilia. At a time of dramatic territorial reconfiguration, when Europeans were intent on discovering, demarcating, and legitimizing new forms of modern national boundaries, the map emerged as an enormously influential medium. The desire, indeed the impulse, to make maps was not restricted to a privileged few. While cartography had once been the domain of a powerful ruling elite, the period between the French Revolution and the Second World War saw an unprecedented proliferation of mapmaking across different ranks of society.[1] This book is about the diverse and innovative approaches to mapping boundaries that arose from this era of popular cartography. It is the story of how cartophilia spread from the gilded map rooms of royal palaces and the gated headquarters of army general staffs, to middle class geographic and historical societies, to humble one-room schoolhouses, transforming how ordinary Europeans viewed the lands around them.

Nowhere was the modern European passion for mapmaking more evident than in Alsace-Lorraine, where the territory's disputed national status inspired a lively dynamic of mapping and counter-mapping.[2] Repeatedly changing hands from the French to the Germans, Alsace-Lorraine became a mapmaker's laboratory, a place subjected to multiple pictorial interpretations and competing topographies.[3] Some mapmakers trapped the borderland within the straight, cage-like lines of scientifically ordered grids. Others used bright colors to highlight the borderland's linguistic geography. Some cartographers appealed to their public's artistic and emotional sensibilities, embellishing their maps with cartouches overflowing with decorative scenes of rural peasantry, local heroes, and native flora and fauna. Still other mapmakers created images that were not what we traditionally think of as maps at all, but rather carefully selected visual surveys of the borderland composed of photographs, lithographs, and drawings. Together, this diverse body of maps reflected a period of tremendous innovation in European cartography, when the people who made maps, the people who looked at maps, and the content of maps themselves underwent revolutionary change.

While the proliferation of mapmaking and map reading across modern European society was certainly made possible by faster and cheaper printing technologies, it was largely motivated by the new kinds of questions that nationalism posed about territory. Before the French Revolution, border mapping was largely a means for rulers in state capitals to secure their military and administrative power over the far-flung reaches of their realms. Using sophisticated optical instruments and angle-measuring machines—the products of Enlightenment science—professional map surveyors imposed the first ordered system of visible, enforceable boundaries on European land. But beginning in the late eighteenth century, European views of territory changed, becoming deeply interconnected with the idea of the nation. Learning how to see European land through the new logic of nationalism was in no way simple or obvious. How exactly does one find the point in territorial space where the French nation ends and the German nation begins? When does a French mountain become a German piece of rock? When does a French village begin to look like a German one? Answering these slippery questions demanded new approaches to mapping borders and border territories, and it is these creative, imaginative, if amateurish, mapping methods that form the heart of this study.

The following pages will demonstrate that a significant number of the cartographers and surveyors who mapped Alsace-Lorraine had no formal training in scientific land surveying, astronomical observation, or trigonometry. Few wore official state uniforms that anyone would recognize. Some came from professional backgrounds in the rising fields of social science, including geographers, ethnographers, anthropologists, linguists, and historians. For these members of the educated bourgeois classes, mapping the social geography of the French-German borderland became a helpful cognitive exercise for making sense of their research on the metrics of nationality. Other popular mapmakers—including hikers, historical preservationists, schoolteachers, and priests—developed emotional and sensory readings of the disputed territory. They designed maps for outdoor use that promoted intimate, firsthand contact with the borderland's mountains, farms, cities, and villages. Despite the variation in their cartographic methods, citizen mapmakers in Alsace-Lorraine shared a similar motivation. For them, mapmaking was about a search for identity; it was about orienting themselves in evolving national and regional communities, both physically and psychologically.

Because they made their maps outside the jurisdiction of state-sanctioned border commissions, however, the innovative work of popular mapmakers is often left out of the story of European border making. The archives of the French-German Border Commission contain just one kind of map: scientifically measured images produced by government-paid surveyors. These maps present the story of an official European border system that became increasingly fixed and stabilized, both on paper and on land, from the eighteenth century onward. But if we broaden our cartographic archive to include popular maps from Alsace and Lorraine, we completely change the story of how European borders were made. We see that during the age of nationalism, the official French-German border coexisted with a significant number of alternative, culturally defined borders that were the work of citizen surveyors. None of these unofficial borders—linguistic, ethnic, historical, even racial borders—carried any legal or jurisdictional weight. All were invisible. But when mapped across paper, they took on an air of popular legitimacy, challenging the authority of state-sanctioned borders and throwing the question of what made a "good border" into public debate.[4]

Maps, in other words, can help us to explain the tensions that developed between real and ideal spaces in modern Europe. While official maps reveal how centralizing European states controlled their border territories through military and bureaucratic channels, unofficial maps can tell us about the future states and idealized borders that European citizens hoped to bring into being.[5] They can also tell us about how informal territorial concepts such as the French "homeland" or the German "hometown" acquired emotional resonance and symbolic power during the nineteenth and twentieth centuries. Printed in large quantities and disseminated broadly across a variety of social circles, unofficial maps had a powerful impact on public opinion. They were masterly communications devices that became objects of desire, enabling nationalist and regionalist groups to stake their territorial claims through their carefully drawn unofficial borderlines and border spaces.[6]

European historians typically characterize the early modern period and the much more recent phase of European integration as eras of fluid borders

and territorial ambiguity. In contrast, the intervening age of European nationalism conjures up images of strict border markers, watchtowers manned
by armed guards, and customs offices crammed with bureaucratic forms.
This book challenges these assumptions by demonstrating that European
land became increasingly malleable during the nineteenth and twentieth
centuries thanks to the democratization of cartography. As more and more
people became interested in making and reading maps in places like Alsace-
Lorraine, the modes of viewing and interpreting European land became
increasingly heterogeneous. The age of European nationalism was thus
hardly a period in which territorial borders became hardened and fixed.
Rather, it was an era of heightened territorial discourse and contestation in
which civilian mapmakers played a critical role.

VISUAL DEVICES THAT INTERPRET AND COMMUNICATE

Telling the story of disputed Alsace-Lorraine through its intriguing, often
enigmatic cartographic archive requires the historian to take a leap of faith
in the validity, and indeed the centrality, of maps to the study of the past.
To open our minds to maps as valuable objects of historical analysis, we
must toss aside the illusion that maps are somehow mirror images of the
places that they represent. All maps are biased views of land.[7] They do not
reflect territory; they interpret it. Like an artist using a paintbrush, mapmakers use a variety of visual techniques to highlight certain aspects of
their subject matter while hiding other features in the shadows. Mapmakers have at their disposal a number of visual tools, including grids, figures,
insets, colors, and toponyms. They can play around with scale, choosing to
zoom in on a rural Alsatian village or to show the village as a tiny dot on
the map of Europe. Or they can make innovative decisions about perspective, choosing to picture the French-German boundary from a bird's aerial
view or from a panoramic vantage point.[8] It is precisely these kinds of visual choices that make maps so valuable as windows into the past; their
biases enable us to discern the cultural and political views of their time.[9]

For many years, historians of cartography used the biases inherent in
maps to uncover the coercive, centralizing power of modern European
states.[10] They stressed how rulers living in European capitals used modern scientific maps to impose visual order and control over their expansive
realms. According to this line of thinking, European maps were "weapons
of the state system" whose all-seeing, panoptic views of land helped states
to crush the independence of European provinces and localities, particularly their remote borderlands.[11] In terms of their psychological influence,
state-sponsored maps "desocialized" territories, both at home and in overseas empires, by reducing the social and cultural complexity of territory to

a coldly calculated system of signs and measurements.[12] While this interpretation of maps has yielded important insight into mechanisms of state
authority and control, it has limited our ability to explore the power and
influence of maps outside the halls of government.

The problem with interpreting modern European maps as top-down
tools for government oppression, exploitation, and control is that it dis-
misses the potential for cartography to serve as a positive form of *self-
identification* for members of a society. This study will argue that maps did
at times serve the authoritative and even violent needs of states to establish
mastery over the French-German border territory. But it will also argue that
border mapping slipped out of the control of government agencies and became an outlet for people from different corners of French and German
society to express their identities, emotions, and loyalties. Historians working outside Europe have already begun to uncover a powerful connection
between popular mapmaking and identity formation. We have learned, for
example, that mapmaking played a key role in the construction of national,
religious, and cultural identities in Russia, India, and the United States.[13]
Rather than emphasize state-generated cartographic surveys that served the
interest of powerful rulers, scholars working in the field of popular cartography have demonstrated how ordinary people used maps to explore their
own beliefs about their territorial surroundings.[14]

In the context of modern Europe, we can think of the democratization
of mapmaking as part of a broader movement toward public engagement
in a variety of scientific fields, including botany, chemistry, astronomy,
and anatomy. During the nineteenth century, professional scientists were
increasingly upstaged by popular scientists, including women, who used
the commercial print market and public forums to disseminate scientific
knowledge to large national audiences.[15] The spread of cartographic knowledge across European society followed a similar pattern. While profession-
ally trained cartographers and surveyors created maps for a handful of elite
institutions, popular cartographers printed amateur maps for their entire
nation to see. The divide between popular and professional mapmakers was
certainly never black-and-white, just as there was always some overlap between popular and professional scientific circles. But the different motives
behind high and low cartography were undeniable.[16] We cannot measure
the full impact of maps on European history without a better understanding of the low cartography that dominated map production from the nineteenth century onward.

To make sense of how cheap, mass-produced, publicly generated maps
shaped territorial identities in modern Europe, we need a methodology for
exploring how maps swayed public opinion and shaped discourses on national belonging. We know that maps can be masterly rhetorical devices. But
analyzing a map's visual language—understanding how a map "talks"—is

only part of understanding its persuasive power. Like newspapers or radio, maps were objects of mass communication in modern Europe. They became part of associational life, public institutions, and commercial ventures. Figuring out how people encountered, read, and used maps is the key to uncovering how they communicated information to their audience.[17] Borrowing methods from the history of the book and media studies, we will investigate how maps circulated in different social settings across Alsace-Lorraine—ranging from classrooms to public exhibitions—in order to understand their psychological influence on the border population.[18]

In exploring how French and German citizens encountered maps in their everyday lives, this study will also pay special attention to how cartography interacted with other forms of mass-produced visual media during the "age of mechanical reproduction."[19] When Europeans came across maps in atlases, magazines, and public exhibitions, they frequently saw them clustered together with other types of visual images, especially lithographs and photographs. The very idea of "mapness" in the nineteenth century was incredibly broad.[20] A series of photos taken along a walking itinerary or a richly illustrated lithograph often supplemented the information contained in more traditional-looking maps. This is why we must never tear old maps away from the other visual images that once surrounded them on paper; if we remove a map from its original context in a tourist guidebook or a classroom geography manual, we lose track of the web of meaning that it formed with the photographs, lithographs, and texts on adjacent pages. When we resituate maps within the larger visual culture of their time, we gain a deeper understanding of how they promoted new mentalities toward territorial space.

MAPPING NATIONS INTO BEING

Maps gave form and meaning to territorial space during a time of rapid change in European politics and culture. As reproducible paper representations of land, they were ideally suited for convincing people to reconsider their older territorial attachments and embrace the new concept of a "national" homeland. One of the first scholars to connect the spread of cartophilia to the success of national movements was Thongchai Winichakul, a historian of modern Thailand. In his innovative book, Thongchai suggested that we think of the nation-state as a "geo-body" whose utopian visual form was more fantasy than reality.[21] Even though the idealized national boundaries that appeared on Thai maps had no official power, he argued, they were instrumental to the spread of Thai nationalism because they made the nation's geo-body appear real, and even historical, in the eyes of the Thai public. It was Thongchai's formative study on Thai nationalism that served as the foundation for Benedict Anderson's analysis of maps in his seminal work on the nation as an "imagined community."[22] Recent studies have expanded upon these Southeast Asian scholars' spatial approach to nation building, demonstrating how simple, mass-produced representations of national space served as powerful "logo maps" for nations in the making across the globe.[23]

But even though existing studies have contributed a great deal to our understanding of how imagined national communities became territorialized through cartography, our definition of what constitutes a "national" map has remained overly simplistic. This study seeks to expand our archive of "national" maps to include thousands of amateur maps produced at the margins of national territory, in the peripheral borderland of Alsace-Lorraine. In shifting the focus away from map production in the capital cities of Paris and Berlin, this study weighs in on a long-standing debate over the spread of national identification across territorial space. For many years, historians argued that nationalism was an ideology that diffused spatially outward from European state capitals. They characterized nation builders as passionate and fearless men who traveled from Europe's geopolitical centers out to the rural countryside, converting backward provincials into national citizens through education, military conscription, and a modern capitalist infrastructure.[24] Over time, however, historians have discounted this centralist approach to the spread of nationalism. Studies on nation building in provinces located far away from state capitals—including the Cerdanya, Brittany, Flanders, Württemberg, and the Palatinate—have demonstrated that the success of European national movements depended on the work of enterprising provincials who strategically adapted national ideals to local circumstances.[25] The prevailing argument among French and German historians today is that nationalism's rise depended on the creation of powerful relationships among local, regional, and national feelings of territorial attachment. Given this consensus view, it is surprising that historians have yet to explore how European mapmakers used different scales and visual techniques to represent territorial space on paper.[26] If we follow the argument that national identification in modern Europe was a decentralized process, then we need to figure out how amateur and local representations of territory fit into the larger story of how people mapped European nations into being.

Many of the citizen mapmakers at work in disputed Alsace and Lorraine were searching for a common ground between their abstract national geo-body and their tangible local surroundings. Instead of adopting a single, one-size-fits-all logo map of their national territory, provincial mapmakers defined the visual limits of their French and German homelands through a variety of different kinds of maps that relied on a creative set of vantage points and decorative imagery. These amateur mapmakers demonstrated, through the diversity of their cartographic representations of land, the many paths toward becoming national in modern Europe. One of the paths toward national identification, for example, went through the subnational space of the region.[27] As Kären Wigen demonstrated in her study on the Nagano Prefecture in Japan, regional maps were valuable resources for citizens seeking to define the relationship between their regional and national homelands.[28] Like Nagano, Alsace and Lorraine appeared frequently on nineteenth- and twentieth-century maps as distinct territorial units. How did these maps of regional space influence how citizens visualized the French and German nations as a whole?

The popularity of regional maps in Alsace and Lorraine was remarkable when we consider the devastating blow that revolutionaries struck against the Old Regime's provincial power structure during the French Revolution. In 1790, one year into the revolution, a new French government attempted to eradicate France's territorial particularities by literally wiping the names of provinces off the official map of France.[29] The revolutionaries replaced territorial spaces like Alsace and Lorraine, which they considered to be dangerous strongholds of noble power and unfair privileges, with "enlightened" administrative departments, named after rivers and mountains, which they designed to promote equality across French territory. But something surprising happened after the revolutionaries forcibly removed France's historic provinces from official view. French citizens resurrected the memory of the obsolete provinces on their own. During the nineteenth century, bourgeois notables from across France formed civil associations with the goal of preserving their regional dialects, histories, cultures, and ecologies.[30] Thanks to their efforts, the regions of Alsace and Lorraine remained alive—and even thrived—as imagined, unofficial, French territorial spaces until the late nineteenth century.[31]

Then, in 1871, the German Empire redrew the boundaries of the French-German borderland by creating a new territory called Alsace-Lorraine out of the lands that it had conquered from France during the Franco-Prussian War. The brand-new region of Alsace-Lorraine was composed of most of the historical region of Alsace and approximately one-third of Lorraine, lands that had never been jointly administered before. Called the *Reichsland*, Alsace-Lorraine was fundamentally different from an old Regime French province because it had the official status of a German federal state, albeit with special rules that kept it under the close watch of imperial authorities.[32] While turn-of-the-century German mapmakers framed Alsace-Lorraine as the newest piece in the puzzle of their federated empire, however, another group of mapmakers refused to think of Alsace-Lorraine as part of German national territory. In this study, I refer to the cartographic work of Alsatian and Lorrainer regionalists as "counter-maps."[33] Rather than accept the inevitability of a Europe broken up into bounded nation-states, regionalists proposed an alternative map of Europe composed of independent, self-sustaining regions. In so doing, Alsace-Lorraine's regionalist mapmakers demonstrated how cartography was a medium that served not only the interest of state power, but also those seeking to resist the dominant national border system of their time and to reimagine space in radically new ways.

It is important to recognize that all of the tensions and philosophical disagreements that unfolded between French, German, and Alsatian mapmakers were part of a much larger European story. Alsace-Lorraine's mixed cultural, linguistic, historical, and religious heritage was typical of many European borderlands. The fact that the majority of Alsatians, along with many of those living in eastern Lorraine, spoke German-based dialects echoed the linguistic geographies of border territories like the Tyrol, Bohemia, Posen, and the Cerdanya, where border populations did not always speak the same language as the majority of citizens in their country.[34] Further, Alsace-Lorraine's complicated past as the territorial possession of multiple states and empires was analogous to the unstable histories of rule in places like Poland, Italy, and the Balkans, where nationalists could also pick and choose among a variety of historical borders to stake territorial claims. My argument thus rests on the premise that Alsace-Lorraine was not a historical anomaly, but a trendsetting laboratory in which European citizens first experimented with innovative kinds of popular cartography that shaped new forms of popular identities.

CARTOGRAPHIC GENRES

This book is organized in a way that highlights the diversity of mapping methods practiced in disputed Alsace and Lorraine. Each chapter focuses on a different type of map that French and German governments, nationalists, and regionalists invented between 1789 and 1940 to advance their territorial claims. To help us understand how different forms of cartography became tools of power for parties involved in the border dispute, I employ both iconological and sociological methods of analysis. First, I examine the maps' visual language: their scale, orientation, perspective, use of color, and symbols. Then, I trace the maps' "social lives" by exploring how people surveyed, printed, distributed, and displayed them. Each genre of map, I argue, promoted its own mode of viewing Alsatian and Lorrainer territory and offered a distinct framework for interpreting the meaning and function of the French-German boundary. When we compare the different kinds of maps with one another, we discover—in stark visual terms—how European border making became an incredibly complex, multilayered, and contested process during the age of nationalism.

While the thematic organization of this book helps to illustrate the variety of mapping genres from which the French and Germans could choose, it also serves another purpose. In writing this book, I have aimed to present a European history of Alsace-Lorraine that breaks down the barriers between French and German national histories. Each chapter will make cross-national comparisons that reveal key differences between French and German philosophies of national boundaries.[35] But each chapter will also take into account transnational transfers in French and German techniques for surveying, printing, and using border maps.[36] French and German nationalists may have disagreed passionately over their claims to Alsace-Lorraine, but they ultimately shared many of the same methods for visualizing their borders.

Chapter 1 begins our journey through Alsace-Lorraine's map archive with an exploration of state-sponsored boundary maps. I begin with state-commissioned maps because they constituted an important first step in the border visualization process and served as references for all of the subsequent kinds of border maps in Alsace-Lorraine. Beginning in the eighteenth century, European states began to "fence in" their territories with precisely measured borderlines. In order to draw and demarcate these modern boundaries, European states created mapping institutions that trained professional corps of surveyors in the use of scientific instruments and triangulation techniques. When they were dispatched to border territories such as Alsace and Lorraine, the state-sponsored surveying corps may as well have been entering terra incognita. They were outsiders who possessed little or no knowledge of the border territories' linguistic, cultural, or historical heritage. Their cartographic vision of Alsace-Lorraine was dictated by the rules of scientific observation and measurement. As a result, French and German surveyors—though they worked for rival states—produced nearly identical maps of disputed Alsace-Lorraine. The two states' cold, gridded, and dehumanized maps of the border territory were so similar, in fact, that they transferred easily between the two sides, particularly in times of war.

The scientific method of border surveying was thus quite restricted in terms of the kinds of people that it involved and the visual form of the maps that it generated. Beginning in the nineteenth century, however, the practice of border mapping spread outside the halls of government-run cartographic institutions and into the homes and meeting places of middle-class European nationalists. In chapter 2, I explore how civilian mapmakers from France and Germany redefined the imperatives of border mapping to reflect the demands of national movements. Many of these popular mapmakers were experts from the rising social scientific fields of ethnography, anthropology, geography, and history. Rather than using scientific survey instruments, they experimented with drawing borderlines between French and German territories based on cultural evidence such as archaeological finds, census records, travelogues, and archival documents. As a result of their efforts, thousands of "unofficial" French-German boundary maps began to circulate in the nineteenth century, challenging the "official" boundary sanctioned by the French and German states. While these alternative boundary maps wielded no legal or jurisdictional authority, they became a powerful means for nationalists in both France and Germany to prove the legitimacy of their claims to Alsace-Lorraine before a public audience of map consumers.

Chapter 3 focuses on one of the most influential kinds of unofficial border maps circulating in nineteenth-century Alsace-Lorraine: the language map. For German nationalists who believed that language was the most important signifier of national identity, language mapping became a favorite method for arguing that Alsace and Lorraine naturally belonged inside a German state. Throughout the nineteenth century, cartographers working on behalf of middle-class German civil associations developed innovative language maps that projected their linguistically defined nation across territorial space. In Alsace-Lorraine, German language maps were met with strong local resistance. Alsatian regionalists rejected binary images of a French-German language border in favor of maps depicting an interconnected network of dialect and subdialect boundaries. Neither French nor German, regionalist-inspired maps laid the groundwork for an alternative European order structured along regional lines.

While the maps that I discuss in chapters 2 and 3 had the power to promote new boundary concepts in the nineteenth and twentieth centuries, they were of limited use in changing how people *experienced* border territories. Highlighting a cerebral set of social scientific "facts," they lacked the visual form and local perspective necessary to connect with the emotions or life experiences of ordinary Alsatians and Lorrainers. Another cartographic genre—"vernacular" or "everyday use" maps—filled this need. Unlike the boundary maps printed for the purposes of elite nationalist discourse, these maps of Alsace-Lorraine were often designed for outdoor use, where they could guide people through a direct, sensory experience of land. Invented by civilian mapmakers, these mass-produced, low-cost maps had the aim of changing how ordinary people in the borderland perceived their everyday surroundings, helping them to forge a psychological connection between their familiar spaces and their national and/or regional identity. The second half of this study focuses on three of these innovative types of maps: village maps, hiking maps, and urban maps.

Each time that the official French-German border moved back and forth between the Vosges and the Rhine, thousands of villages found themselves in a new national community. Chapter 4 analyzes how mass-produced village maps became powerful pedagogical tools for teaching Alsatians and Lorrainers about the relationship between their local hometown and their national or regional homeland. Printed in small Alsatian publishing houses, local maps—particularly cadastral and classroom maps—presented images of land on a scale that was detailed enough for villagers, with their limited travel experience, to recognize from experience. Using embellished decorative imagery that defied rational modes of territorial representation, local mapmakers transformed villages into utopian places. They promoted an idealized understanding of nationally rooted villages that fitted within a constellation of other "French" or "German" hometowns. Like language maps, however, village maps could be appropriated by regionalists as tools of cultural resistance. Rather than situate the borderland's villages within a larger map image of France or Germany, Alsatian regionalists created their own counter-maps that attached their villages to a self-sustaining regional space.

While some cartographers focused on remapping Alsace-Lorraine's picturesque villages through the prism of national or regional identity, others developed innovative ways of visualizing the borderland's natural environment. In chapter 5, I focus on the French, German, and Alsatian civil associations that created the first series of hiking maps for the Vosges Mountains. Promoting nationalized and regionalized views of the Vosges through hiking maps was far from simple. It was a loosely defined process in which civil associations first selected landmarks of national and regional interest in the Vosges, then created a pictorial or photographic archive of the landmarks, and finally developed hiking maps to introduce their visual archive to the greater public. Foldable, cheap, and designed to be carried into the wilderness, hiking maps distinguished themselves from other kinds of cartographic media in disputed Alsace-Lorraine because they were specifically designed to define border space through the movement of bodies. Like the other types of maps designed for everyday use, hiking maps frequently appeared alongside lithographs, photographs, and other forms of visual media that captured the firsthand experience of moving through the land. Examining the relationships between these different kinds of territorial images helps us to understand the connection between map- and landscape making during the age of European nationalism.

In the last chapter of the book, I turn from the cartographic construction of mountain landscapes to the invention of border cityscapes. Located directly on the Rhine border between France and Germany, Strasbourg was the administrative and cultural capital of Alsace-Lorraine. During the era of competitive nationalism, manipulating the appearance of Strasbourg's cityscape (*image de la ville*, or *Stadtbild*) became increasingly important for the French and the Germans who were seeking to legitimize their cultural claims to the borderland. Both national groups attempted to leave a lasting physical mark on the city in the form of new buildings, neighborhoods, and ports. In addition to their new works of architecture, nationalists and regionalists printed their own city maps for sale in local bookshops. The French, German, and Alsatian maps of Strasbourg each guided their readers through a distinct visual experience of the city.

Compared to the beautiful, handcrafted manuscript maps of Europe's medieval past, the cartographic images that we will encounter in the following pages can appear rough and crude. But what these modern, mass-produced, ephemeral maps lost in beauty, they gained in power. Thanks to their widespread availability, the power to see land from a commanding bird's-eye position, and manipulate how it looked, became accessible to ordinary Europeans for the first time. In Alsace-Lorraine, we will see maps stained with the blood of French and German soldiers, thrown out the back of trains, drawn on with crayons, and carried into the wilderness. They will be ripped, traced, filed away, folded, sweated through, dirtied, and replaced. By the twentieth century, we shall see that cartography was a medium of mass communication that had inserted itself powerfully into the public arena, helping to transform border making from a matter of government policy into a subject of lively public debate.

PART I

Mapping Borders

CHAPTER 1

States Map Their Borders

Before the invention of modern maps, it was impossible to visualize what European territories looked like from above. It was only thanks to the scientific advances of the early modern period that cartographers learned to construct an aerial view of land from measurements taken on the ground. New understandings of the rotating heavenly bodies—the stars, the sun, and the moon—and the development of intricate angle-measuring machines laid the foundation for a modern visual perspective on European territory based on rationality and order rather than on imagination, spirituality, or artistry. With their gridded matrices, carefully measured distances, and meridians oriented toward state capitals, modern survey maps dramatically changed how Europeans looked at their borders and border territories. Beginning in the eighteenth century, places that had once felt remote and culturally estranged from European capitals became visible, and even knowable, because of maps. By collapsing the psychological distance between center and periphery, they transformed how European rulers exercised power over their people, enabling them to "see" and manage their distant border regions without the physical need to travel.[1]

Scientific survey maps are the oldest form of maps that this book will examine, and they will also be the most restricted in terms of the people that sponsored their creation and participated in their production. States alone could marshal the financial resources and manpower necessary to survey vast expanses of European land with newly invented surveying methods and technologies. While many of the forms of mapmaking practiced in Alsace and Lorraine were open to amateur cartographers, the "official" map surveys of France and Germany (the *Carte d'État-Major* and the *Generalstabskarte*) were under the domain of professionalized corps of surveyors. These surveyors did not visualize the French-German borderland in the same way that local inhabitants did, as a familiar place with its own particular linguistic, historical, and cultural topography. Rather, the French and German governments trained their surveyors to see and codify the borderland according to the "universal" rules of modern science. The geographies that rival French and German states created for Alsace and Lorraine were thus imperial views of land commissioned from afar, characterized by a highly regulated, scientifically prescribed and standardized form of visual representation.[2] The borderline separating modern France and Germany likewise became a testament to modern scientific prowess, demanding hundreds of hours of painstaking precision measurements in riverbeds and thick forests, places that defied easy mapping.

For the people living in Alsace and Lorraine, the scientific, totalizing view of land dealt a crushing blow to their local autonomy. The very idea of a map survey, which has the same root as the word "surveillance," can be defined as "the act of looking at something as a whole, or from a commanding position."[3] Beginning in the eighteenth century, centralized, government-run cartographic institutions in Paris, and later Berlin, assembled, processed, organized, and printed vast amounts of territorial knowledge about the border regions in the interest of controlling everything from taxation to canalization to roads. But the greatest demand for scientifically surveyed boundary maps came from the French and German militaries. So great was the militaries' need to perfect their "battle vision" that official state map surveys fell under the direct supervision of the French military from 1793 to 1940 and under the militaries of the German states from approximately 1816 to 1921. By creating synoptic views of land that established a mental distance between map reader and territory, survey cartography made it possible for the French and German states to exert power over Alsace and Lorraine in increasingly brutal, efficient, and mechanized ways.[4]

Comparing how the French and the Germans managed their cartographic surveys of Alsace and Lorraine can reveal a great deal about the particular relationships between territorial knowledge and power, and centers and peripheries, in the two countries.[5] But such a comparison does not only illuminate national differences; it also uncovers the remarkable degree of cooperation between European states that bordered each other.[6] The paradox was that as the modern French-German border became more clearly defined, both on paper and on land, transnational flows of cartographic knowledge increased. This is because the formal separation of French and German territory demanded a great deal of coordination between the two states. Some of these transnational exchanges took the peaceable form of mailing scientific articles, purchasing instruments, and collaborating on binational border commissions. But a great deal of the cartographic knowledge transfers took place under the pressures of war and occupation. The story of the production and circulation of border maps in Alsace-Lorraine thus provides new evidence of the historical interdependence or "entangledness" of European states, even during the time of their greatest conflict.[7]

LOCAL BEGINNINGS

The scientific approach to mapping European land originated in the sixteenth century, when European sovereigns first pursued cartography as a tool to protect and administer their realms during a period of increased competition for resources and territory.[8] The earliest map of Alsace that responded to the heightened defensive needs of European states was commissioned by the Habsburg monarchy, which at the time still ruled the border region from Vienna. First printed in 1576, the *Elsasskarte* was a transitional map that reflected both a lingering interest in humanist aesthetics and a forward-looking military concern for topographical detail. Its author was Daniel Specklin, a Strasbourg-based artist and fortification specialist. Specklin's experience of mapping Alsace would differ fundamentally from the experiences of the French and German state-sponsored surveyors that followed him. The Habsburg commission would mark the last time that an Alsatian was put in charge of surveying the border region, and it would mark the last time that an individual would undertake the task of mapping Alsace on his own. Specklin's *Elsasskarte* thus serves as an enduring point of contrast to later French and German survey maps of Alsace that were imagined and executed in imperial fashion from the center out to the periphery.

It took Specklin three years, from 1573 to 1576, to complete his map, which would be printed in separate sections for Upper and Lower Alsace.[9] He surveyed the map with little assistance from others, crisscrossing the region on foot, noting astronomical observations, measuring distances by counting his paces, and sketching the map in the field. Drawing on his experience in military defense, Specklin conceived the map according to a "cavalier perspective,"[10] a term used by fortification builders to describe the experience of looking down onto land from an elevated defensive position. As a result, the Rhine River, the Alsatian Plain, and the Vosges Mountains appear stacked on top of one another. In spite of the inaccuracies resulting from its cavalier perspective, Specklin's map was groundbreaking in the scope and exactitude of the settlements that he represented. The scale of the map, 1:190,000, was dramatically larger than those of previous maps of the region, on the scale of 1:500,000 or 1:700,000. It was thus able to provide a detailed view of land favored by administrators and generals in

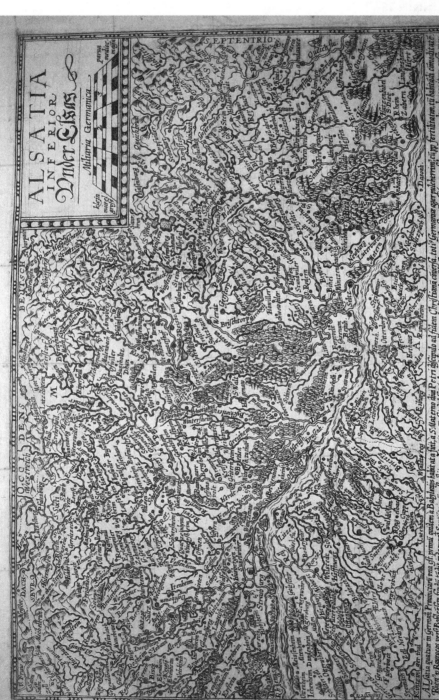

Figure 1.1 Daniel Specklin's map of *Alsatia Inferior* (*Under Elsas*), commissioned under the decentralized Holy Roman Empire. Drawn according to cavalier perspective, it shows the peaks of the Vosges Mountains at the top of the image and the Rhine River in the bottom third. Photo. et coll. BNU Strasbourg, M.Carte.100.011.

Vienna. A remarkable achievement for its time, Specklin's map served as an important military reference for the Habsburgs during the Thirty Years War from 1618 to 1648, when Alsace became a battlefield for warring European states.[11] At the same time, the map's high-quality drawing work demonstrated the fortification builder's more traditional artisanal training: as a young man, Specklin, the son of a Protestant tradesman in Strasbourg, had been apprenticed as a silk embroiderer.

Combining local and imperial perspectives on Alsatian territory, Specklin's *Elsasskarte* reflected the decentralized power structure of the Holy Roman Empire. On the one hand, the image confirmed that Alsace was the imperial possession of a faraway emperor in Vienna. The symbol of a Habsburg eagle in a circle denoted the status of some Alsatian settlements as "imperial cities." Distances between localities were recorded with the standard imperial measurement system, the *Miliaria Germanica*. Moreover, the text inscribed at the base of the map sheets celebrated the Germanic roots of Alsace. Employing a classic humanist style, the text began: "Alsace is one of the four provinces in the German land. It was first settled by the Babylonians, and converted to the Christian faith by Saints Matthew and Peter. It is the most beautiful area of Germania, due to its fertileness and habitations."[12] Alongside these affirmations of imperial control, however, Specklin labeled his map with toponyms that suggested a locally rooted cultural perspective. Specklin recorded town, abbey, and castle names in the dialect spoken by the Alsatian population rather than in High German: Ropenum instead of Roppenheim, Blessen instead of Bläsheim, and Mittelwihr instead of Mittelweier.[13] Even though Specklin's map of Alsace was commissioned by a sovereign seeking to consolidate territorial knowledge of his empire, the map maintained a certain fidelity to regional culture. In the coming years, European states would eliminate this hybrid model of a local/imperial map in favor of a triangulation network created by centralized cartographic institutions and a system of uniform topographical symbols.

Specklin's map also offers an informative contrast to later maps of Alsace that used the Rhine River as a semiotic indicator for the rupture between French and German territories. His inclusion of a number of right-Rhine communities into the body of his map (see the bottom third of fig. 1.1) reflected the economic, social, and cultural realities of life in the Rhine region during the sixteenth century. During this period, before certain French philosophers and government officials claimed the Rhine as France's "natural border,"[14] the waterway existed as a significant point of friendly contact between Alsatians and Badeners. Enlightenment ideas, Protestant beliefs, and commercial goods flowed easily between the right and left banks of the Rhine. The Germanic dialects spoken on either side of the river were nearly identical, allowing for easy verbal communication. It is perhaps not surprising that the Germans who annexed Alsace from

France in 1871 sought to revive the mental world encompassed by Specklin's sixteenth-century map, recreating the kinds of trade and cultural exchanges in the Rhine area that they believed to be the historical and "natural" orientation of Alsace, toward the east.

BORDER MAKING BECOMES A SCIENCE

The Kingdom of France seized parts of Alsace from the Holy Roman Empire in 1648 (it annexed Strasbourg in 1681), at around the same time that French administrators decided to make cartography a state priority. When Jean-Baptiste Colbert, the chief minister of Louis XIV, founded the Royal Academy of Sciences in 1666, one of his foremost goals was to establish a system of cartographic surveys that would produce accurate two-dimensional images of the kingdom. To lead this immense state-sponsored enterprise, Colbert chose the Italian astronomer Jean-Dominique Cassini (Cassini I). For the next century, Cassini I and his descendants (Cassini II–IV) lived at the Observatory of Paris and produced an innovative set of maps that "emphasized scientific principles over the pictoral tradition,"[15] replacing the culturally rooted image of the French kingdom as the "body of the king" with an abstract grid. The close collaboration between state and science— the royal patronage of mathematicians and astronomers led by the Cassinis, the state-funded construction of an astronomical observatory, and the state's ability to use coercion and compromise in the provinces—helped France to become the leader in European cartography during the seventeenth and eighteenth centuries.[16]

The Cassinis completed their first map of France, the map of the "Great Triangles," in 1744.[17] The goal of the map was to create a sweeping aerial view of the French kingdom. To do so meant that older methods of orienting maps, such as Daniel Specklin's "cavalier perspective," would be abandoned in favor of an innovative way of measuring and representing territory. This new method, called triangulation, was a process of spatial reasoning designed to eliminate the distorted territorial views of an individual observer by covering a piece of land with a network of thousands of imaginary triangles.[18] To draw the triangles, the French academy commissioned a team of young, low-paid surveyors, called *ingénieurs-géographes*, to travel across the kingdom and establish some two-thousand triangulation points that would form the "skeleton" of the map. In Alsace, Cassini III's team used a variety of triangulation points on the landscape, including church steeples, wind and water mills, castle towers, and farm pillars.[19] The surveyors then measured the length of one side of each imaginary triangle by dragging chains between signal points. Last, they used trigonometry to calculate the lengths of the other two sides of each triangle. The result was

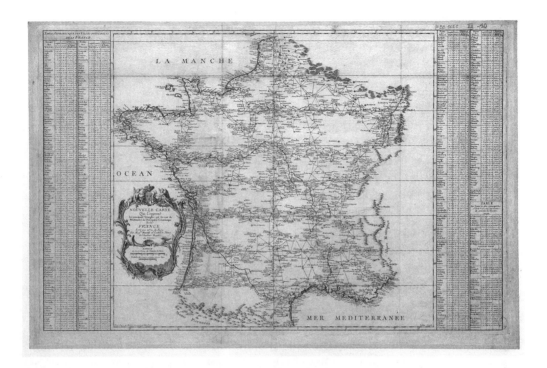

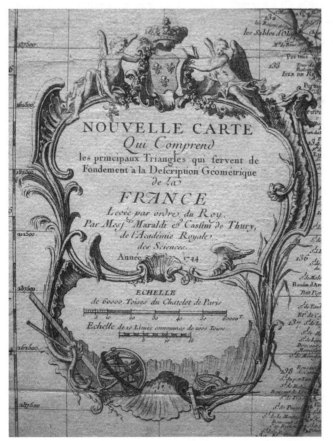

Figure 1.2a Cassini III, *Nouvelle carte qui comprend les principaux triangles qui servent de fondement à la description géométrique de la France,* 1744. The French kings led Europe in creating scientific surveys of their realm using modern triangulation techniques. Note the chains of triangles stretching across the map image. Bibliothèque nationale de France, Ge BB 565 A (VII), plate 10.

Figure 1.2b The marriage between science and royal power in eighteenth-century France appears in the symbols on the Cassini map's cartouche. At the top, cherubs hold the fleur-de-lis crest of France's Bourbon monarchy, while surveying instruments lie at the bottom. None of the individual surveyors who worked on the map receive acknowledgement. Bibliothèque nationale de France, Ge BB 565 A (VII), plate 10.

an image of France produced by "scientific reason," a great accomplishment during the age of enlightenment.[20]

Indeed, the imperatives of eighteenth-century French cartography reflected the visual equivalent of what Enlightenment philosophers desired from government: transparency, order, and rationality. One of the most striking features of the Cassini map was its support for the idea of an "enlightened" state that was highly centralized, confirming the often-cited link between the art of surveying and the art of governing.[21] Within the map image, the Observatory of Paris appeared as a beacon from the capital, the place around which the rest of France oriented itself. Though it was clearly selected because of its proximity to the center of political power, the observatory's location took on the air of scientific legitimacy, establishing the baseline measurements for north, south, east, and west, and therefore the geometric coordinates for longitude and latitude. The distance charts pictured to the left and right of the map image further reinforced the idea of a France oriented spatially toward Paris: the Alsatian capital of Strasbourg, for example, is listed at a distance of 205,269 toises from the Parisian observatory. A close-up image of the map (see plate 1) shows the meridian of Paris slicing through Alsace, binding the region to the center of France along its axis and symbolizing the noose that a "well-ordered" centralizing government was wrapping around the neck of its recently conquered borderland.

But how tightly was the noose wound? Though the Cassini family presented its maps as "objective," scientifically reasoned representations of France, the images were in fact highly utopian visions of French territory. Their views of a seamless, integrated French state papered over the realities of the kingdom's enduring cultural and linguistic diversity. Up through the French Revolution, Alsatians continued to practice a unique set of local laws and hold church services in their German-sounding dialect. Even the scientific knowledge that underpinned the Cassini maps had been won at great price. Though some locals had helped the Cassini surveyors to perform their land measurements, there were also significant cases of resistance to the surveyors in both Alsace and across France; in spite of the king's orders, some inhabitants, including priests, refused the surveyors access to observation towers like church steeples.[22] On the final Cassini map, however, the power struggles between center and periphery disappeared quietly into a commanding view of a unified French kingdom.

In addition to bringing order and homogeneity to the visual image of the state, French surveyors began to transform French boundaries into scientifically fixed lines. On the map of the "Great Triangles," the boundary between France and Germany remained quite ill defined. It was not until the waning years of the Old Regime that the French government made a concerted effort to formally "fence in" its peripheral territories.[23] During the

1770s and 1780s, the French monarchy charged the military surveyor Jean-Claude Le Michaud d'Arçon with the task of expanding the Cassinis' land survey to include some three hundred detailed field survey sheets (1:14,000) of France's eastern border along the midpoint of the Rhine River from the Jura to northern Alsace, called the *Carte des frontières de l'est* (see plate 2).[24] The on-the-ground reality of demarcating the border, however, proved to be enormously challenging. Mapping the boundary line soon turned into a nascent canalization project; the only way to create the border was to stabilize the Rhine's constantly shifting river topography. D'Arçon's desire to construct a more "rational" border was visible in the smooth channels that he drew alongside the river's "natural" path, its multiple fingers colored blue with a disordered jumble of islands in brown. When political geography did not match well with natural geography in eighteenth-century France, maps provided a utopian space for making a border appear clean and orderly.

The utopian messages present in the French kingdom's maps of its borders and borderlands did not, however, outweigh the maps' important practical uses. Once the surveyors returned to Paris, the maps were engraved, printed and circulated, becoming "immutable mobiles"—identical mass-produced images—which officials from different administrative and military branches of government could consult when they made decisions about Alsace and Lorraine, which became part of France in 1766.[25] But even beyond the institutions of the state, the maps found an eager audience in the French public. Cassini III himself explained: "As much as it is advantageous for a sovereign to know the countries under his rule, it is also useful for his subjects to know the location of places that their interests or their business may lead them."[26] Though the "public" that purchased the state's survey maps remained mostly limited to elites, the eighteenth century marked the beginnings of a much greater popular interest in French geography. The notion that maps should serve the public good became especially important to French revolutionaries, who were quick to declare the Cassini maps, along with the Old Regime's castles and churches, a form of "national property" (*bien national*). The revolutionary Committee of Public Safety even created its own outfits for state surveyors, emblazoning their buttons with fasces and the words "Surveyor of the Republic."[27] But France's radical revolutionaries also made a lasting structural change to the management of the state map survey, transferring control over the production and distribution of maps from the Academy of Sciences to the French Army in 1793. The state's geographic knowledge was soon put to use for Napoleon's far-reaching military conquests, and the Cassinis' enlightened tool for "reasoning terrain" would encourage a militarized view of land that would turn natural environments and human settlements into battlefields.

THE MILITARIZATION OF STATE MAPMAKING AND THE
TRANSNATIONAL EXCHANGE OF CARTOGRAPHIC TECHNIQUES

While the French Army had possessed its own corps of surveyors since the late seventeenth century, its new role as the overseer of state cartography after 1793 ushered in significant changes to how French people visualized their land. As Enlightenment-era scientists interested in triangulation and trigonometry, the Cassinis had neglected to map the three-dimensional aspects of French terrain: changes in vegetation such as fields, forests, and marshes, or the rise and fall of mountains and hills. French revolutionary military commanders, on the other hand, placed a high value on the shape and height of land, essential knowledge for any army on the move. It was Napoleon Bonaparte, France's great military leader, who first called for an ambitious new map survey of France with an emphasis on topographical forms, called the *Carte d'État-Major* (General Staff map). This map would be drawn by military surveyors directly in the field, with the help of portable drawing boards and levels, devices used to approximate the height of geographic features. To represent their new territorial knowledge on paper, French military cartographers also developed an innovative pictorial language for representing topographical relief.[28]

German rulers would benefit directly from the newly militarized form of French cartography. Immediately following the Peace of Lunéville in 1801, French surveyors set to work establishing topographical surveys of the German lands that Napoleon's armies now occupied. The kings and princes that presided over these lands were generally supportive of the conquering army's mapping initiatives because they provided the rulers with a coveted modern tool for administration and tax collection. The French and Germans agreed upon a mutually beneficial plan. German rulers would provide the necessary financing, matériel, and surveyors to establish maps of their lands, and the French occupying forces would lend the Germans expertise and assistance for surveying, drafting, and engraving the maps. Before the Napoleonic invasion, the German states had made some headway in mapping their territories, but they lacked serious, methodical cartography on the scope of France's Cassini maps. The day-to-day encounter between German surveyors and French military personnel enabled German states to make significant improvements in their maps within a relatively short time.

For the Royal Prussian Land Survey, which would later be responsible for mapping Alsace-Lorraine, the collaboration with Napoleon's surveyors was a formative experience.[29] French colonel and surveyor Jean-Joseph Tranchot, for example, was instrumental in leading German teams to conduct land surveys in Prussia from 1803 to 1813. But surveys of Prussian lands were also led by Germans employing French methods. Locally trained surveyors, including the future head of the Prussian Land Survey, Friedrich Karl Ferdinand von Müffling, used French measurement techniques to produce a Westfalian map identical in scale (1:86,400) and paper size to the map of France. In 1815, after the Wars of Liberation, von Müffling drew on his experiences with Napoleon's surveyors to complete the Prussian army's instruction manual for survey maps, establishing an official Prussian state "cartographic language" for representing land relief.[30] The manual would remain a fundamental reference text for German surveyors during the next thirty years. Even the language used in the text reflects the close collaboration between French and German surveyors: the German mapping term *Cotirung*, meaning slope, was derived from the French *côte*, while *nivellement*, the French term used to describe the leveling technique employed to measure the height of hills and mountains, simply became *Nivellement* in German.[31] Contacts between French and German surveying teams during the Napoleonic period also had a lasting effect on the institutional culture surrounding mapmaking in Germany. The cartographic bureaus of several powerful German states—Prussia, Bavaria, and Württemberg—were all founded under Napoleonic rule.

When Napoleon's European empire collapsed, German states were thus well positioned to carry out their own systematic map surveys of the new boundaries established by peace treaties in 1814–15. When the boundary between France and the German lands officially returned to its prerevolutionary location at the midpoint of the Rhine River, the grand duke of Baden commissioned a set of detailed, scientifically surveyed maps of the border, completed in 1828 and measured in *badische Ruthe*, the equivalent of three meters.[32] The German duke agreed with French authorities that a corps of surveyors from Baden would be responsible for the border measurements on the right bank of the Rhine, while a corps of French surveyors would be responsible for measurements on the left bank of the Rhine. As their surveying work proceeded, the heads of the two surveying missions communicated regularly with one another.[33] The final set of maps, collectively called the *Topographical Map of the Rhine and Its Two Banks*, was printed on a large scale (1:20,000) in order to demonstrate the scientific legitimacy of a state boundary measured with no fewer than 581 trigonometric signal points and 1,390 triangles. While only a few decades earlier, the French surveying corps had been the most advanced in Europe, the Germans had caught up quickly, becoming equal partners in mapping the new French-German border after Napoleon's demise.

It is important to recognize, however, that the grand duke of Baden's carefully measured Rhine boundary from 1828 did not necessarily translate into a stronger psychological divide between the French and the German

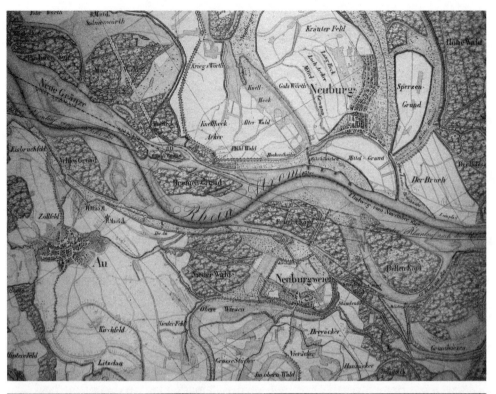

Figure 1.3a The grand duke of Baden commissioned his own set of topographical maps of the Rhine border following the Napoleonic Wars. Note the label for the new border, or *Neue Gränze*, in the upper left-hand corner of the image. *Topographische Karte des Rheinstromes und seiner beiderseitigen Ufer.* Bibliothèque nationale de France, Ge CC 27 42.

Figure 1.3b On the frontispiece of the *Topographische Karte des Rheinstromes*, female figures representing the two sides of the Rhine gaze peacefully at one another as ships glide along the waters in the background. Bibliothèque nationale de France, Ge CC 27 42.

people who lived along the river. On the frontispiece for the duke's map, two female figures, each representing a different bank of the river, look peacefully toward one another. Strasbourg and its cathedral spire are visible behind the female on the left, while the mountains of Baden's Black Forest are visible behind the female on the right. Pictured in the waters on both sides of the Rhine are boats, representing the promise of international commerce in the early nineteenth century. Indeed, the map's completion coincided with the enormous cross-border project of canalizing the Rhine and transforming it into a profitable waterway for industrial business leaders.[34] Understanding the Rhine's flow and mapping its minute topographical details was thus not always motivated by the desire to create a "fence" between France and Germany; this particular border map represented the desire to create a peaceable cross-border commercial culture.

But the mutually beneficial cross-border relationship embodied in the visual language of the map's frontispiece would never become a lasting reality. In 1870, when German armies crossed the Rhine to attack France during the Franco-Prussian War, they relied on detailed survey maps to help them, even as the French tried in vain to hide their maps of the region surrounding Strasbourg.[35] By the time that the short but enormously consequential war was over, the boundary between France and the German lands would need to be redrawn again. This time, however, French and German military surveyors would not carry out their work separately, but would instead collaborate on their first-ever joint border commission.

DRAWING THE NEW BORDER AFTER 1871

The impressive advances in the science of mapmaking that took place during the peaceable mid-nineteenth century placed maps in a position to play a critical role in treaty negotiations following the Franco-Prussian War. In September of 1870, even before the war had come to an end, the Royal Prussian Land Survey circulated a map of a proposed new border between France and Germany, indicated in green, which planned the transfer of the French departments of the Haut-Rhin, the Bas-Rhin, and the new Department of the Moselle[36] to the German Empire, creating a new territorial entity called the *Reichsland* or Alsace-Lorraine (for a satirical perspective on the power of war victors to carve up territory, see Otto von Bismarck taking a pair of scissors to the French-German border area in fig. 1.4). The map provided close-ups of different sections of the new boundary, showing how the line would be drawn through local settlements and geography.[37] On 10 May 1871, "the map with the green line,"[38] signed by Jules Favre and Otto von Bismarck, was officially enshrined in Article I of the Treaty of Frankfurt as

Karikatur des Kladderadatsch.

Figure 1.4 A political cartoon from a popular German magazine pictures the German chancellor, Otto von Bismarck, cutting Alsace-Lorraine off the map of France. *Kladderadatsch,* March 1871. Author's collection.

the visual reference for the new French-German border. Because the new limit of sovereignty between France and Germany was set on land, roughly along the midpoint of the Vosges Mountains, and not along the midpoint of the Rhine River, where it had previously stood, this map served as an essential set of instructions for how to demarcate the "man-made" border. To ensure that border markers would be placed according to the precise letter of the law, the Treaty of Frankfurt called for an international commission to execute the layout of the new frontier.[39] This international commission, with equal numbers of French and German members, met shortly after the conclusion of the peace conference to begin the tedious task of transcribing the green line from the two-dimensional treaty map onto a three-dimensional mountain terrain.

Over the course of their joint meetings, the Germans and French drew up a set of instructions for the marking, measurement, mapping, and description of the French-German border in the Vosges Mountains. They divided the 506-kilometer-long border into ten sections, each under the authority of one German and one French surveyor.[40] Together, these teams of German and French surveyors performed careful measurements and made trigonometric calculations to lay down border markers with as much fidelity as possible to the treaty map. From Luxembourg to the Swiss border, the international commission laid down fifty-five hundred stone border markers (*bornes* or *Grenzsteine*), each costing 26 francs and 20 centimes,[41] displaying a "D" for "Deutschland" on one side and an "F" for "France" on the other. During the 1870s, the commission established a series of large-scale (1:20,000) maps of the boundary. As an official legal document, each map bore the stamp of the German and French chief geometricians in charge of the expedition.[42] Not unlike the joint work of the French and German surveyors during the Napoleonic occupation decades earlier, the stamps of the French and German geometricians represented the shared cartographic standards of the two European powers, and their mutual faith in the "universal" application of scientific reason for solving international disputes. Negotiators from both states placed such trust in the viability of

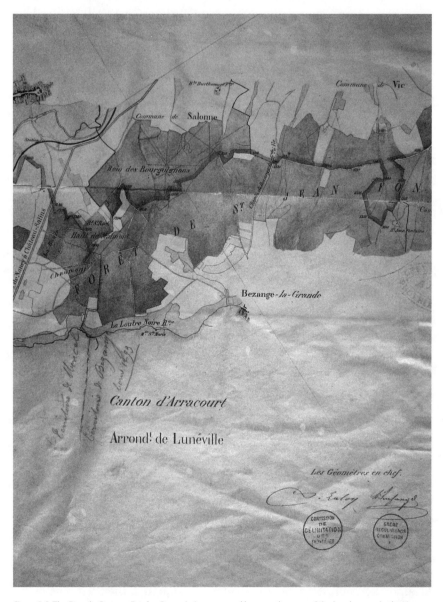

Figure 1.5 The French-German Border Commission surveyed large-scale maps of the border area in the Vosges Mountains. Note the stamps of approval from the heads of the French and German surveying teams in the bottom right corner of this map from 1871. Archives départementales de Meurthe-et-Moselle, 1 Fi 21. D.R.

their boundary maps that they agreed to give them legal precedence over the written description of the boundary.[43]

For the border populations whose lives were turned upside down by the treaty settlement of 1871, the work of the French-German Border Commission was met with resentment. Not only was the commission in charge of demarcating and mapping the new border, but it was also assigned a secondary task: it was responsible for splitting the goods and capital of any communities traversed by the new boundary between the German and French states. For the communities of Arraye and Ajoncourt, for example, the new border disrupted a close and mutually beneficial relationship.[44] Before 1871, the tiny mountain communities had shared a fire station, a cemetery, and a church. The Treaty of Frankfurt determined that the new border would divide the two communities: Ajoncourt would become German and Arraye would remain French. The French-German Border Commission ordered an inventory of the material wealth in both communities, and the communal religious property was split, down to the last candlestick. The priest for the communities of Arraye and Ajoncourt wrote letters of protest, pointing to the villages' shared "spiritual community." [45] His letters, however, had no impact on the border regime that authorities in Berlin and Paris were constructing in the Vosges. The states' border-mapping methods reinforced the idea of a "desocialized" border landscape that disregarded the complex social dynamics on the ground in Alsace-Lorraine in favor of bureaucratic efficiency.

Despite protests from the local population in the Vosges Mountains, from 1871 to 1914 French and German authorities worked hard to ensure that the official border between their two states remained faithful to treaty agreements. Every year, the international commission checked on the status of the border markers and consulted with maps to ensure that the legal limit of state sovereignty between France and Germany was accurate and clearly visible.[46] The task was a frustrating one. Members of the commission complained that the local population moved the stones at their convenience. One border marker was found cemented into a farmhouse whose owner claimed that he had not noticed that the stone he retrieved from a field had any special markings on it. Another set of markers was moved by the local population to build a bridge. Other reports pointed to political motivations. Officials interpreted the removal or destruction of border markers as a sign of protest against the German status of Alsace-Lorraine.[47] These signals from a local population who overlooked—or actively protested—the new border illustrated the tensions that arise when states' legal division of territory overturns strong social and economic networks at the local level.[48] Meanwhile, as pro-French groups were left to mourn the "blue line of the Vosges," the German surveyors had a much happier task at hand: creating the first map of the German Empire.

A. Arnould, édit., Mars-la-Tour

La Frontière entre Mars-la-Tour et Vionville

A l'occasion de l'anniversaire de la bataille du 16 Août 1870, il se célèbre chaque année à Mars-la-Tour un service pour nos soldats morts au champ d'honneur. Nombreux sont les visiteurs des deux nations. Des Gendarmes Français et Allemands gardent la frontière.

Figure 1.6 Postcard picturing French and German soldiers in front of a border marker at Mars-la-Tour. Author's collection.

THE VISUAL UNIFICATION OF GERMANY

The achievement of German unification on the political level in 1871 was followed by a visual unification of German territory on paper. Just as the Cassini maps had answered to the demands of an increasingly centralized French state, the German imperial government set to work on incorporating each diverse corner of its empire into a single cartographic template. "One of the essential needs of every civilized state [*Kulturstaat*]," declared the head of the Royal Prussian Land Survey, "is the possession of accurate and good maps."[49] Unlike the Parisian authorities that managed the French national map, however, the army institutions responsible for mapping the German Empire operated within a federal structure. A number of the German states had a long and respected history of mapmaking, in particular the cartographic bureaus attached to the Army general staffs of Bavaria, Baden, Württemberg, Saxony, Hesse, and Prussia. In 1878, members of these six topographical bureaus gathered in Berlin and agreed to collaborate on the first *Karte des Deutschen Reiches*, or *Map of the German Empire*.[50]

The military convention of 4 March 1878 established that the German imperial map would be modeled on the existing Prussian and Saxon maps on the scale of 1:100,000, a scale small enough to provide a commanding view of the entire empire but large enough to illuminate topographical detail. In addition to this map, the convention provided for the establishment of a *Generalstabskarte* (General Staff map) of the empire consisting of nearly four thousand *Messtischblätter* (small plane-table surveys) on the scale of 1:25,000, for close-up views of localities. Demonstrating the similarity between the art of ruling and the art of surveying, German federal politics found their way into the orientation of the German imperial map's regional components. Prussian maps would use the Berlin Observatory as their meridian, while Bavaria, Saxony, and Württemberg would continue to refer to their own observatories until 1910. French maps, as we recall, referred exclusively to the Parisian Observatory. While the surveying process was relatively autonomous for each of the six imperial German cartographic bureaus, the resulting imperial map nonetheless achieved the visual unity that German nationalists sought. Each parcel of German land shared an identical *Zeichenerklärung*, or graphic code: a unified German set of normative symbols for representing railroads, streets, bodies of water, land formations, towns, and cities (see plate 3).[51] The "cartographic language" of the federated German state was therefore just as uniform as it was for the French republic.

Situated on the westernmost periphery of the German Empire, Alsace-Lorraine was an area of particular concern for Emperor Wilhelm I and his advisers. Even though the region was nominally the collective possession of all twenty-five German states, Bismarck insisted that topographical surveys of Alsace-Lorraine fall under the domain of Prussia alone.[52] This was perhaps because the newly reorganized Royal Prussian Land Survey, a self-standing institution within the Army General Staff, had the largest group of personnel out of all the German lands (258 employees in 1875), divided into trigonometric, topographical, and mapping sections with a total annual budget of 1,026,700 marks.[53] But the move perhaps also reflected the Berlin administration's desire to oversee military defense directly vis-à-vis France. It is telling that the earliest Prussian land surveys in Alsace-Lorraine focused on the Franco-Prussian War battlefields near Metz, Woerth, Weissenburg, and Spichern.[54]

A commemorative map in honor of Emperor Wilhelm illustrated the German Empire's militarized perspective on Alsatian territory.[55] The *Erinnerungsblatt* from 1886 pictured a topographical survey map of Alsace framed by the wood of an oak tree, one of the symbols of the German nation. The adult Prussian ruler appeared at the top of the map, while miniature portraits of the emperor as a youth in military costume framed the left and right sides. The bottom panel referred to the military exercises that troops from the Fifteenth Regiment, garrisoned in Alsace-Lorraine, had conducted in honor of the emperor. The decorative framing of the topographical survey map suggested Emperor Wilhelm's domination over German national

Figure 1.7 Commemorative map of military exercises conducted in honor of Emperor Wilhelm's visit to Alsace-Lorraine in 1886. The map image is framed by the wood of an oak tree, one of the symbols of the German nation, as well as scenes from the young emperor's life. *Erinnerungsblatt des Kaisermanövers des XV. Armee Corps 1886*. Photo. et coll. BNU Strasbourg, M.Carte.10202.

territory and created a militarized visual image of the borderland as the national war prize from 1871.

In addition to revealing its concern for military security, Prussia's cartographic projects in Alsace-Lorraine reflected a strategy for transitioning the territory's infrastructure and economy from French to German rule. "The Royal Land Survey has two aims," explained one Prussian surveying publication. "It should produce a unified map of the state territory for military purposes as well as for technical and state economic projects."[56] The more detailed topographical information that the German administrators in Berlin and Strasbourg could attain, the better informed they were of recent changes to the Alsatian landscape resulting from industrialization and urbanization:

> The need for a topographical map of Alsace-Lorraine that corresponds to the current relationships of the territory has become increasingly apparent over time. The sheets of the great *Map of France* (1:80,000) that cover this area

are for the most part over forty years old and have only been slightly corrected as of late. However, in recent years a fundamental rearrangement has already occurred in the *Reichsland*, including the reestablishment of numerous German place-names and considerable improvements to the railroad and street networks, such that the older maps are no longer useful for specialized work.[57]

The numerous requests that newly arrived German bureaucrats in Strasbourg issued to the Royal Prussian Land Survey demonstrated the high demand for maps on behalf of civil servants who had to collect taxes, organize transportation, and operate police forces in an unfamiliar, and sometimes hostile, territory.[58]

While the Royal Prussian Land Survey's maps made important updates to preexisting French maps of Alsace-Lorraine in terms of railroad, transportation, and communications networks, it is striking that the trigono-

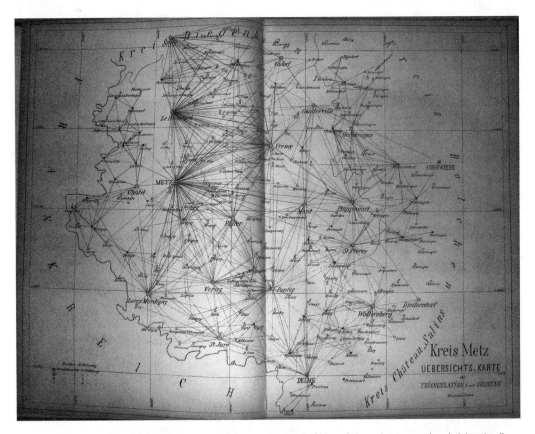

Figure 1.8 Prussian surveyors relied on a vast network of triangulation points to map the administrative districts in Alsace-Lorraine. A number of these points were first established under French rule, an illustration of the transnational exchange of cartographic knowledge in nineteenth-century Europe. Map of *Kreis Metz* from *Uebersichtskarten der Triangulation 3. & 4. Ordnung (1:100,000)*. Archives départementales du Bas-Rhin, 4K10.

Fig. 2.

Trigonometrisches Signal auf einem Punkt I. O.

Figure 1.9 When Prussian surveyors could not locate a point high enough to conduct their trigonometric measurements, they built their own signal towers, ten to thirty meters tall, with an observation platform on top. The towers stood as a testament to the modern European state's desire to acquire clear, unobstructed views of the territory under its control. *Die Arbeiten der Königlich Preussischen Landes-Aufnahme* (Berlin: Königlich Preussischen Landesaufnahme, 1893), figure II. Reproduced with permission from the Staatsbibliothek zu Berlin–Preussischer Kulturbesitz.

metric and topographical content of the maps did not change significantly from the French period. Because Germans used land measurement techniques similar to those of French surveyors, they could easily incorporate French cartographic knowledge of Alsace-Lorraine directly into their own maps. In fact, the first German map of the border region, published in 1879, was called the *Map of Alsace-Lorraine Based on Special Reconnaissance using the "Map of France."*[59] During their subsequent four-stage retriangulation of Alsace-Lorraine, begun in 1879, Prussian surveyors built directly upon the older French network of cartographic measurements, using preexisting French triangulation points as the basis for their revised maps.[60]

While these updated maps served as important tools for German functionaries and military authorities—Berlin's "hands and feet" in the Alsace-Lorraine—it was also during the period of the German annexation that ordinary Alsatians and Lorrainers first acquired widespread access to topographical survey maps. Correspondence between Berlin and Strasbourg emphasized the importance of selling the maps at a low cost (one mark per sheet) to make them affordable for use by private individuals and businesses.[61] The Royal Prussian Land Survey achieved these low costs thanks to advances in printing technology. By the late nineteenth century, lithography had replaced copper plates in the map production process, which permitted the images to be engraved directly onto stone. While the final product was less fine and precise than maps printed from copper plates, the benefit was that maps could be printed more quickly and cheaply. Business contracts between bookstores in Alsace-Lorraine and the Land Survey headquarters in Berlin helped to facilitate the commercial availability of the images. As a result, ordinary people in Alsace and Lorraine began to purchase Land Survey maps for a variety of their own purposes, often using them as templates to produce new maps for their historical and beautification societies, hiking clubs, classrooms, and tourist associations. The blossoming civilian map industry in Alsace-Lorraine would suffer a severe blow, however, from the military's ban on the public circulation of maps during World War I.

BATTLE VISION

Under the pressures of World War I, government and military leaders in Paris and Berlin carefully censored the print media available to the public. Both the French and the German states determined that topographical survey maps provided visual information about the national territory that was too sensitive for public use in wartime, asserting their monopoly over space, as well as time.[62] In Alsace-Lorraine, preparations for the confiscation of commercial maps began in 1910.[63] The head of the German army corps stationed in Alsace-Lorraine called for a list of local businesses that sold maps or owned printing stones for their publication. He attached a list of the Alsatian maps of particular concern, including all of the maps produced by the Royal Prussian Land Survey, as well as maps produced by civilian associations and tourist agencies. In 1912, when the army first informed the public that it would carry out its planned map confiscations through a newspaper article in the *Strassburger Zeitung*, the Association of Alsace-Lorrainer Booksellers sent letters of protest to the German imperial authorities. "The *Reichsland* bookstores have for decades tried to assist the public in getting to know maps," one letter complained. "The population has for a long time been accustomed to coming to bookstores for their map needs."[64] While the bookstore owners were of course acting according to their economic self-interest in defending their freedom to sell maps, their protests nonetheless raised the question of the status of maps as a "public good" in German society. Did the military have a right to confiscate maps from ordinary German citizens? Had the public's growing passion for buying and reading maps become too much of a threat to the all-seeing eye of the state? The answer was clearly yes.[65]

During World War I, map use in Alsace-Lorraine became synonymous with the demands of "battle vision"—knowing where to move, where to shoot, where to launch projectiles, and where to retreat.[66] Along with written reports and telephone calls, maps were a key form of communication regarding the status of battlefield activities: "As soon as the operations began, the role of maps became critical in the correspondence exchanged between the different echelons of command and those carrying out orders, between the troops of different armies. . . . No precise indication could be given without the help of a map: it constituted the keystone of the liaison."[67] The maps' wartime liaison route, it turned out, bypassed ordinary soldiers. Army reports indicated that company commanders had access to maps, while the man in the trenches had no choice but to look blindly to his company commander to guide him through unfamiliar terrain.[68] Along with cigarettes and food, maps became a popular item on the soldiers' black market. In the German Army, a map could fetch a price of ten to twenty marks.[69]

Over the course of the war, the French and German armies perfected the art of battlefield mapmaking, producing extremely detailed maps on the scale of 1:10,000 or 1:5,000 (called *plans directeurs* by the French and *Messtischblätter* by the Germans). Army reconnaissance units reprinted the maps several times a week in order to give military commanders the latest territorial views of the battlefield based on new intelligence from neighboring armies, declarations from prisoners, and documents seized from the enemy.[70] The interchangeability of French and German survey maps had an important strategic impact on the war. When a muddied paper trail of maps lay scattered across kilometers of battlefields, army units were sent to recover enemy maps from cadavers, a fact which is immediately noticeable to the historian in the dirt and bloodstains covering the captured German maps at the French Military Archives.[71] French army units used recovered German *Messtischblätter* (1:25,000) of Alsace-Lorraine for their operations in the Vosges Mountains, converting the German maps into *plans directeurs* to direct artillery fire at the battles of the Linge and Hartmannswillerkopf (see plate 4).[72] French and German armies did their best to ensure that their armies used a homogenous set of conventional signs to clearly communicate battery positions, enemy supply lines, and other key indicators for "battle vision." Over the course of the war, they were able to figure out each others' conventional signs through intelligence so that they could use captured maps. The scale of map production on both sides was enormous. Eventually, the lack of paper on their home front forced the German Army to stop printing large-format maps and switch to a smaller size.[73] The French Army, which had printed just 6,000 *plans directeurs* in 1914, printed 4,460,000 in 1918 with the assistance of mobile map-printing trains and printing stations, including one in Gugenheim, a village in Upper-Alsace.[74]

The usages of topographical survey maps during World War I reflected a militarized vision of landscape that originated from the scientific ordering of terrain that began in the Enlightenment.[75] In wartime, the visual translation of natural environments, villages, and towns onto numerated grids encouraged a desocialized mentality toward battlefield terrain that enabled armies to inflict destruction on a piece of land without an awareness of the human beings who lived there. Maps guided what armies refer to as the "organization of terrain"—the placement of forces and equipment, the digging of trenches, and the establishment of supply chains. A passage from a postwar Michelin guidebook described how the French Army prepared the mountainous Vosges terrain for the Battle of Morhange in German Lorraine, near Metz, during the earliest days of the war in August 1914: "The crests had been organized in great secret since the first of August in a remarkable manner: concrete trenches, preceded by networks of iron cables and scattered machine guns; in the back, heavy artillery and light artillery were housed. The terrain had in advance been repaired, measured, and detailed down to its smallest recesses. Maps with grids were at the disposal of the artillery, the infantry, and the aviators."[76] Of course, the French Army's goal of slowly increasing its possession of Alsatian territory during World War I proved disastrous for those unfortunate villages caught in the middle of the war of position. The Alsatian village of Steinbach, for example, located near the famous battlefield of Hartmannswillerkopf, was identified simply by its position at *Côte 425* (Slope 425), a numerical indicator taken from a military survey map of the area. A postwar Michelin guide recounts how the 152nd French Army Division had to destroy the village because of its vital strategic location: "At the cost of incredible sacrifices and efforts, the 152 at last reached Steinbach. The struggle then became savage. In the middle of the fires and uninterrupted gunfire—coming from the basement windows of cellars, from roofs, from crenellated walls—there had to be a siege of every house."[77] The Michelin guide contains an image of the "martyred village" of Steinbach with a line marking its key strategic position at *Côte 425*, a mountain promontory offering a highly desirable view of German troop movements in the Alsatian Plain below.[78] Steinbach itself had been reduced to ruins. The "reasoning" of terrain through scientific measurements thus supported a military approach in which villages could be objectified as numbers and commanders could give the strategic value of land a greater value than the lives and livelihoods of the people who lived on the land.

GERMAN MEASUREMENTS IN THE OFFICIAL MAP OF FRANCE

After 1918, the democratic postwar German government transferred control of the national map survey to a civilian agency based in Berlin, called the Reichsamt für Landesaufnahme, bringing an end to nearly one hundred years of military-led mapping in the German Empire. In France, on the other hand, the national map survey remained under military control. Immediately following the armistice, the Geographical Service of the French Army set to work on creating a French-language map of Alsace-Lorraine that Parisian administrators urgently needed to transition the region to French rule. To create the new map as quickly and as efficiently as possible, French officials maintained the Prussian Army's topographical and triangulation data for Alsace-Lorraine, as well as the conventional symbols and scale of the German *Generalstabskarte* (General Staff map) of 1:100,000. Following the example of the Germans in 1871, Parisian administrators and military personnel relied heavily on preexisting maps of the borderland, referring to the German maps of Alsace-Lorraine as "trustworthy documents."[79] French officials also negotiated with their German counterparts for detailed descriptions of their four triangulations of the region from the turn of the century. The French government commissioned Colonel Paul Poliacchi, the retired head of the Drawing, Engraving, and Foreign Cartography Department for the Geographical Service of the French Army, to create new maps from these preexisting German documents.

Rather than altering the topographical information in the German maps of Alsace-Lorraine, Poliacchi focused on changing the maps' German-language toponyms to French ones and recording changes to what the French call *planimetrie*: communications routes, perimeters of woods that could have been modified during the war, post offices, telegraph and telephone networks, and newly added departmental, canton, and municipal borders.[80] None of these modifications to the German maps demanded the work of surveyors: changes to toponyms required archival research into pre-1871 place-names, and civilian *agents-voyeurs* working for the Vicinal Service noted any changes to *planimetrie* on pre-1914 German maps. It was not until 1922 that the French Army undertook a new topographical survey of Alsace-Lorraine to reincorporate the borderland into the existing 2,835 printing stones that made up the matrices of the official Map of France. Even the French Army's fresh triangulation work incorporated "the preexisting network established by the Prussian Land Survey, situated north of the parallel from Paris."[81]

Maps of Alsace-Lorraine continued to preoccupy the Geographical Service of the French Army during the 1930s, when tensions rose once more between France and Germany. The army redoubled its efforts to survey France's northeastern frontier, where its militarized view of terrain culminated in the Maginot Line. In contrast, the French Army ignored the map of the Spanish border, which had not been revised since 1900.[82] Assuming that a second armed conflict with Germany would be a ground war, the French Army gave the same meticulous attention to the "organization of terrain" in eastern France during the 1930s that it had in the First World War. One image that the Geographical Service produced in 1931 showed how the army was already preparing secret map-printing facilities to mir-

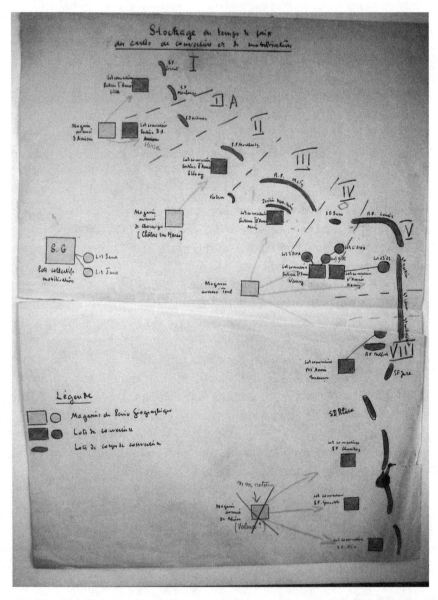

Figure 1.10 During the 1930s, the French Army prepared secret map-printing facilities to mirror a future front against Germany. The curved lines represent France's northern and eastern borders, the squares represent map-printing facilities, and the roman numerals represent French army divisions. *Stockage en temps de paix des cartes de couverture et de la mobilisation*, Service historique de la défense, Vincennes, 9 N 500.

ror a potential second front in Alsace-Lorraine (fig. 1.10). The peacetime storage of maps prepared for army regiments headed to new fronts in Lorraine or Alsace (numbered I–VIII on the image) reflected the lessons learned from the strategic value of maps during World War I. The maps destined for the armies garrisoned on the northeastern frontier during the 1930s included information that was not available to civilians or other government agencies: the location of active and passive forces, portions of routes or railroads considered secret and therefore erased from commercial maps, a telephone network, and depositories of munitions and materiel.[83] Though a well-conceived plan for an earlier war, the maps were of course of little use for the kind of battle that the Germans had planned for 1940.

In spite of the rising German threat, it was during the interwar years that the Geographical Service of the French Army received its sharpest criticism for monopolizing the production and sale of national maps. Proposals to create the civilian-run Institut géographique national (IGN), currently responsible for printing official maps of France, originated in the early 1930s out of recognition that the army was no longer the foremost user of maps—other government bureaus and civilians were now frequent purchasers—and the army did not possess the best marketing skills to reach out to the public. The Geographical Service attempted to convince consumers of their maps' superiority to commercial tourist maps by emphasizing the accuracy and precision of their topographical representations. But the army's interwar tourist maps never achieved the success of privately produced guidebooks such as the Michelin. In 1932, the Geographical Service began to discuss the possible transfer of a large part of the army's mapping services to the civilian sector. But the union representing the army's topographical and cartographic services successfully lobbied to maintain its monopoly on printing state maps, arguing that there was still a need for specialists trained in the "science of cartography."[84] It was not until 1940, under the pressures of the German occupation of France, that the Geographical Service of the French Army closed down and the civilian National Geographical Institute, designed to serve the map needs of a nonmilitary public, opened its doors.

Beginning in the eighteenth century, French and German states developed an innovative visual perspective on their land that emphasized precision, order, and scientific reason. This particular visual perspective, made possible by scientific progress, answered to the growing desires of European states to transform their complex landscapes, composed of diverse natural environments and human settlements, into an easily legible overview. Both the French and the Germans shared the same fundamental understanding of a "good" state map as one that effectively papered over internal differences to promote a cohesive space for the state to exert its sovereignty. The willing erasure of provincial cultures contributed to a "desocialized" view of territory that, at its most extreme, led to battle vision—an emotionally detached perspective on territory that turned inhabited landscapes into battlefields for killing. Of all the places that European rulers needed to map, borders and borderlands were the most important. It was in places like Alsace and Lorraine that European states confirmed their sovereignty over territory by maximizing their ability to "see" and defend the limits of their realms.

The views of Alsace and Lorraine that modern French and German states desired were so similar that they were in fact interchangeable. This was evident in the joint French-German border commissions that shared the same maps, or in the wartime intelligence operations aimed at stealing bloodstained and mud-splattered maps from fallen enemy soldiers. But state-sponsored survey maps were an exception within the broader corpus of maps representing disputed Alsace and Lorraine. The overwhelming majority of maps that pictured the border regions took on a visual form that was not easily interchangeable between the French and the Germans. This is because their purpose was to present map readers with a vision of the borderland's *cultural* topography, and on this subject the French and the Germans did not see eye to eye. As national movements swept across Europe, members of the French and German publics used maps to make cultural arguments for why their national group deserved rights to Alsace and Lorraine. These culturally based maps owed a great deal to the template provided by the states' original topographical surveys. Paris and Berlin may have held exclusive authority over the mapping and demarcation of state boundaries in Alsace-Lorraine, but members of the public could transform the state-produced maps, adding new graphics, color, and language. It is to these publicly generated maps, and the national borders that they helped to construct, that we now turn.

CHAPTER 2

What Makes a Good Border?

Thanks to their sophisticated scientific instruments and rigorous field-work, military surveyors transformed the French-German borderland from an open space into a divided landscape with stone border markers, armed guards, and customs offices. But the story of imagining, drawing, and de-marcating the boundary between modern France and Germany extends well beyond the official state-sanctioned border. During the nineteenth century, new kinds of borders rose to power in modern Europe that would soon rival the boundaries that state cartographic bureaus had so meticu-lously surveyed. This unofficial network of borders, invented by civilian mapmakers, served the function of orienting, framing, and bounding na-tional territory according to cultural, rather than legal, parameters. For na-tionalists, finding the "right" borders to separate their national territory from that of their neighbors became an act of patriotism and an essential part of being a good citizen.[1] As a result of their creative methodologies, citizen mapmakers succeeded in establishing a network of shadow borders that crept over images of modern Europe, fracturing the land into distinct national spaces occupied by different cultural groups. The desire to align Europe's state and national geographies—the dueling products of official and unofficial border surveys—explains how peripheral territories such as Alsace-Lorraine became hotly contested borderlands during the age of nationalism.

Criticizing the "artificiality" of the legal border between France and Germany, citizen mapmakers in Alsace and Lorraine referred to the na-tional boundaries that they invented as "natural," "authentic," and "true." Why did they consider their national boundaries to be better than the of-ficial borders that states had carefully demarcated with the latest scientific technology? In considering its elusive visual form, scholars have referred to the modern nation-state as everything from a "geo-body," to a "mythical space," to an "imaginative geography."[2] Benedict Anderson, the influential theorist of nationalism, considers the map, along with the census and the museum, to be one of the three pillars of the modern nation-state.[3] But we know surprisingly little about the actual cartographic methods that citizens used to map the utopian limits of their national communities across space; we often forget about the hard mapping work that went into "producing" national territories in the same way that scientific surveying technologies "produced" state territories. Who were the citizen surveyors who created the idealized territorial space of the modern European nation? What was their fieldwork like? How did they turn cultural differences into a border system? What kind of relationship did they have with state governments? To answer these questions, we must look outside the official archives of the French-German Border Commission and turn to a variety of archives, printed sources, and popular images that tell us about what the French-German border meant to nationalists, and how they shared their idealized visions for that border with a growing audience of map readers.

The majority of national boundary mappers in Alsace and Lorraine were scholars from the nineteenth century's budding social scientific disciplines: ethnographers, anthropologists, linguists, statisticians, geographers, and historians who were deeply interested in the social and cultural identity of land. Both at home in Europe and abroad in their colonies, this group of French and German middle-class experts in the "science of nationality" set to work on classifying the world's land and peoples into black-and-white national categories based on what they considered to be objective scientific criteria. Some of these researchers made observations and gathered cultural knowledge in the living landscape that they believed could serve as fixed indicators of national identity: geographic landmarks (rivers and mountains), cultural signs (language, clothing, and customs), and buried history (archaeological and anthropological finds).[4] Others turned to quantitative methods, particularly population censuses, to map demographic information across territorial space. Thanks to their innovative research methods, modern Europe's patriotic social scientists succeeded in creating an entirely new genre of boundary map based on the supposed geographic location of national divides. Because each of their boundary maps focused on a single subject of thematic interest, historians of cartography often refer to the visual products of nineteenth-century social scientists as "thematic maps."[5]

Who sponsored this new genre of thematic border cartography? When we examine the biographies of citizen boundary mappers in Alsace and Lorraine, we see that many had ties to state institutions. A number of the social-scientists-turned-mapmakers held academic positions at public universities—in places like Strasbourg, Nancy, Paris, Tübingen, Frankfurt, and

Berlin—where they had access to government statistical and census information, as well as financial support to conduct their research on boundaries.[6] But if we take a closer look at the cartographic archive, we see that many of these social scientists designed their thematic maps for a broad public audience that extended well beyond the academy. Documentation about the sponsorship, sale, and circulation of thematic boundary maps from disputed Alsace-Lorraine reveals that a flourishing network of civil associations—geographic and historical societies in particular—were instrumental to their success. This chapter thus stresses the importance of civil society—networks of individual citizens acting outside the state—to the visual construction of bounded national territories in modern France and Germany.

While thematic boundary maps wielded no legal or jurisdictional authority, patriotic civil associations used them as powerful argumentation tools to sway French and German public opinion on the "Alsace-Lorraine Question." Mass produced, they were available for purchase on the commercial market and appeared frequently in public forums such as exhibitions, classrooms, and meetinghouses.[7] The images of national territory that the maps presented were schematic and idealistic, reducing Alsace and Lorraine's complex cultural topography into simple black-and-white divides. In this regard, thematic boundary maps created similar types of detached territorial perspectives to state-sponsored scientific surveys; they offered synoptic views from above that did not seek to incorporate local perspectives or account for hybrid identities. This is what made them such excellent propaganda tools. Some thematic maps were designed to influence politics at key moments of government transition, such as the revolutions of 1789 and 1848 or German unification in 1871. Others assuaged feelings of guilt and inadequacy after the severing of territory following a catastrophic military defeat, as in France after 1871 or Germany after 1918. The public cartographic archive thus demonstrates how the idea of a "good" boundary map—one that distinguished unambiguously and accurately between "French" and "German" territories—evolved over time and became heavily influenced by national definitions of territorial space.

IDEALIZING NATURAL BORDERS IN
EIGHTEENTH-CENTURY FRANCE

Two conceptual breakthroughs in eighteenth-century France shaped the European public's future role in mapping national borders. First, it was during the eighteenth century that the question of territorial borders definitively broke free from the gilded halls of the Royal Palace, where decisions of where to draw borders had been confined to intimate discussions among

French sovereigns, military authorities, and their European counterparts. Borders entered into public conversation and became the idealized limits of "the nation" thanks to Enlightenment philosophers and other members of the French elite who brought the subject of France's territorial boundaries into their "public sphere."[8] Second, the rise of nationalism in eighteenth-century France succeeded in shifting the bases for border delineations from traditional considerations of international strategy, military defense, and quid pro quo territorial exchanges to "higher" standards of goodness and virtue. For Enlightenment philosophers, the borders that met these high standards had to be stable, enduring, and free from the corruption of human society: they were the national borders that nature had created.

In eighteenth-century France, the concept of a "natural border," as opposed to a man-made state border, grew out of an Enlightenment belief in the power and authority of nature. Elite members of an increasingly influential public, the philosophes would discuss among themselves the meaning of nature as an embodiment of ideals such as truth, virtue, and honesty. In seeming to provide a geographically stable border system that promised not to change over time, the natural limits of rivers and mountains offered this eighteenth-century elite a welcome alternative to the rampant territorial conquests and violence that had marked the reign of Louis XIV. "The lie of the mountains, sea, and rivers which serve as boundaries of the various nations which people it, seems to have fixed forever in number and size," wrote Jean-Jacques Rousseau in 1756. "We may fairly say that the political order of the continent is in some sense the work of nature."[9] The potentially universal application of natural landmarks as border markers separating human societies likewise reflected an "enlightened" worldview; rivers and mountains were common features dividing landscapes across the globe, offering a coherent system of visual signposts for delineating borders both in Europe and abroad. French cartographers from Paris to the New World reinforced the philosophes' preference for natural borders by emphasizing rivers, and particularly mountains, in the decorative boundary iconography of their eighteenth-century survey maps.[10]

Alsace presented one of the key locations for equating natural borders with national borders during the eighteenth century because the midpoint of the Rhine River marked the eastern boundary of both the region and French territory. While the argument that the Rhine River was France's legitimate eastern border dated to Roman times (Caesar famously proclaimed the boundaries of Gaul to be the Rhine, the Pyrenees, and the ocean), the philosophes focused on the natural basis for the border rather than on its precedent as an ancient political border.[11] Thanks to the "natural borders" idea, eighteenth-century French border cartographers went out of their way to use rivers, streams, and mountains to determine France's regional and national limits, even if it meant sacrificing jurisdictional sovereignty and trading border villages with neighboring states.[12] The problem that map surveyors quickly encountered when working with a natural border as large and unwieldy as the Rhine, however, was the waterway's continual mobility. The challenge of mapping the Rhine border in Alsace, detailed in chapter 1, proved illusory the idea that nature's visual symbols offered a clear, cohesive, rational system of national borders between European peoples.

Even if it did not lend itself to easy mapping, the Rhine's symbolic role as the unofficial spatial limit of the French national community enjoyed popularity among the French public well after the fall of the Old Regime in 1789. During the French Revolution, the belief that France's rightful borders were signaled by nature took on new significance for Alsace as French soldiers clashed with the Austrian and Prussian armies just over the Rhine border. Revolutionaries used the concept of natural borders, which had reinforced arguments for peace in the previous decade, as an excuse to redraw the map of Europe by force. Claiming not only the Alsatian part of the Rhine, but the entire left bank of the river as the manifest destiny of the French Republic, the revolutionary leader Georges Danton proclaimed: "[The Republic's] limits are set by nature. We will reach those limits at the horizon, at the banks of the Rhine, at the shores of the ocean, and at the Pyrenees and the Alps."[13] By fixing their eyes on the Rhine as the natural border of French national territory, the revolutionaries became imperialists who used nature as moral justification for conquest.

An image printed in year IV of the French Revolution illustrates the increasing visibility of borders in public discourses over French national identity and imperial expansion into Central Europe (fig. 2.1).[14] The image portrays two angels fluttering above the Rhine River, presenting a map of the border to Marianne, the female symbol of France, who is holding fasces topped with a Phrygian cap (both republican revolutionary symbols), and Germania, who is wearing fur robes and carrying a shield emblazoned with a Prussian eagle. In surrounding the map image of the unofficial border with classical allegories of virtue in the form of cherubs, the image embodies an emerging public narrative of borders as semisacred dividing lines endowed with an air of morality, destiny, and virtue. All the while, a winged victory angel trumpeting in the sky reminds the viewer that France had attained its natural borders through the sacrifices of war.

Accompanying the allegorical image was a text by Georges-Guillaume Boehmer, a German merchant from the left bank of the Rhine and a republican ex-deputy from the national Rhine-German National Convention, who offered a reward of six thousand francs for the citizen who gave the best answer to the question "Is it in the interest of the French Republic to pull back its borders to the banks of the Rhine?"[15] A Parisian publisher printed fifty-six responses to Boehmer's question from persons of all walks of life. "Citizen" Pierre Gadolle, a distiller from Paris, for example, wrote

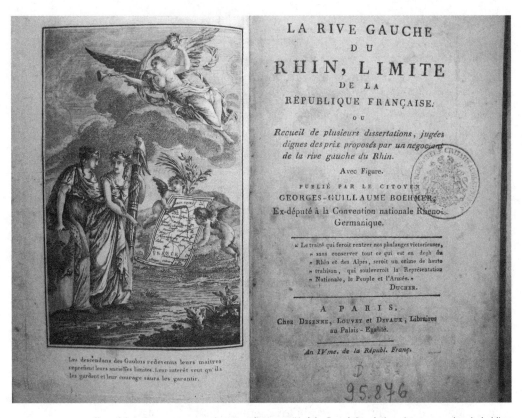

Figure 2.1 An image accompanying a text from year IV of the French Revolution pictures two cherubs holding up a map of France's "natural borders" with Marianne and Germania standing nearby. Georges-Guillaume Boehmer, *La rive gauche du Rhin, limite de la République française, ou Recueil de plusieurs dissertations dignes des prix proposés par un négociant de la rive gauche du Rhin* (Paris: Chez Desenne, Louvet et Devaux, an IV). Fonds patrimonial des médiathèques de Strasbourg.

that "the great rivers and high mountains always make the natural limits and the most secure ramparts of a nation."[16] Boehmer's pamphlet was an early sign that public interest in national borders had the potential to expand beyond French society's educated elites to include ordinary citizens such as a Parisian distiller, who now felt that he, too, had a stake in telling the French state where to map the limits of its realm. While government officials still had the sole legal authority to decide how and where to map the official borders of France, French citizens now *debated* the moral authority of those borders from below.

DELINEATING CULTURAL BOUNDARIES

It was also during the late eighteenth and early nineteenth centuries that German nationalists, disillusioned by the harsh realities of the revolutionary and Napoleonic occupations, began to develop their own definition of a good national border. While they agreed with the French that a legitimate national border should be a natural one, they disagreed that three-dimensional geographic features such as rivers and mountains were the best indicators of a nation's territorial limits. Instead, Germans interpreted the idea of a natural border to mean a line of cultural demarcation between human populations based on barriers of language, ethnicity, or history. Rendering this type of natural border visible on paper demanded innovative social scientific methods. It required the expertise and hard work of skilled German researchers willing to go out into the field and search for the elusive boundary lines that marked the limits of their coalescing national community.

Well before the birth of the German imperial state in 1871, German cultural experts set to work on mapping Germany's national boundaries through field research, surveys, and censuses. The Brothers Grimm became some of the first "national researchers" in nineteenth-century German society to create snapshots of a unified German territorial space, publishing a volume of fairy tales and folklore in 1812 that they had gathered from peasants and villagers from a broad swath of German-speaking Central Europe.[17] But the tales of the Brothers Grimm and other Romantic-era scholars based in Germany's esteemed universities could not provide the kind of geographically pinpointed research that could aid in a formal topographical delineation of a German national border. It was only in the years preceding the revolution of 1848 that German researchers, with the assistance of map imagery, formally identified on paper the territorially based cultural divisions between the German people and other Europeans.

Some of the first culturally based national borders to be mapped in Europe were recorded in Heinrich Berghaus's *Physical Atlas*, published in 1848.[18] After serving in the Prussian Army as a military surveyor during the Wars of Liberation against Napoleon, Berghaus became one of Germany's most influential nineteenth-century geographers and a founding member of Germany's oldest geographic association, the Geographical Society of Berlin. Throughout Europe, geographic societies became an important public forum for members of the educated elite to gather and exchange knowledge about the natural environment and human cultures across the globe.[19] Working closely with maps, geographers also played a major role in shaping European politics, helping to anchor national groups in territorial space through their research on the physical environment, but also on the types of cultures and societies present in European lands.

Like his circle of geographer friends, which included the intrepid explorer Alexander von Humboldt, with whom he collaborated on his *Physical Atlas*, Berghaus was intent on designing an ordered, rational way of organizing current knowledge about the natural world. Berghaus's atlas reflected this categorizing impulse and offered a visual ordering of the world analogous to the mostly text-based French *Encyclopedia* of the eighteenth century. Like encyclopedias, atlases such has Berghaus's became important objects of bourgeois status for members of German society, conferring a level of educational cultivation, or *Bildung*, upon their owners.[20] They were very different objects from state-sponsored survey maps; printed on quality paper with borders hand painted by female workers,[21] they were commercial objects of luxury accessible to a few. Their particular format offered a political opportunity for German nationalists during the 1840s: atlases could replace the dissatisfying status quo legal borders of mid-nineteenth century Europe with new, utopian national borders drawn according to cultural criteria.[22]

Like the French philosophes, Berghaus was fascinated by the connections between nature and politics. Rather than search for Germany's national borders in the environmental topography of rivers and mountains, however, Berghaus looked to "natural" geographic distinctions between people. In placing his culturally based maps inside a physical atlas that contained "general features of mineral and organic nature," the *Physical Atlas* explained the ethnographic divisions between people as part of the natural order of things.[23] The lengthy volume divided knowledge of the world into four categories: plant geography, animal geography, anthropography, and ethnography. The fourth section of the atlas, the "General Ethnographic Atlas" (Allgemeiner ethnographischer Atlas) used ethnographic criteria to render visible on paper the "natural" divisions between human cultures across the globe. Several maps specifically addressed Europe's ethnic makeup, including the *Overview of Europe with Ethnographic Boundaries of Individual States and the Distribution of Peoples in the Middle of the Nineteenth Century*, completed in 1846 (see plate 5).[24] The map's expansive scale, offering a commanding overview of European territory, showed that the image's purpose was argumentative rather than practical. The map visually cleared the way for a new perspective on European territories that was rooted in cultural geography; within the space of the map, ethnic boundaries rivaled existing legal divisions of state sovereignty.[25]

But what exactly did Berghaus mean by ethnic boundaries? We find the answer in an inset that Berghaus called the "Table of Peoples," or *Völker-Tafel*. The inset served as the map's legend, linking the cartographic representation of Europe to an ordered system of ethnographic knowledge.[26] Like the mathematical measurements and scales that appeared in the legends of topographical survey maps, the ethnographic categories in Berghaus's map provided the viewer with a methodological understanding of its own construction. During the mid-nineteenth century, "ethnicity" was still a fluid term that invoked some combination of behavioral and physical signs of culture.[27] To delineate ethnic borders across his map of Europe, Berghaus relied on a combination of two criteria: language and race. Racial categories (organized into the "Table of Peoples") included Celts, Germans, Roman Peoples, and Slavs, whose geographic borders were located according to information that nineteenth-century archeologists had unearthed from excavations.[28] Berghaus then broke down each of these racial types into language groupings. Unlike the racial genealogy aspect of his research, the linguistic aspect was informed by contemporary field studies that enabled him to freeze-frame Europe's ethnic territories as they stood in the middle of the nineteenth century. Hence the phrase in the map's title: "Distribution of Peoples in the Middle of the Nineteenth Century" (Völker-Sitzen in der Mitte des 19'ten Jahrhunderts).

But how long would the ethnic borders that Berghaus delineated stay where they were on the map? Like the Rhine River, which frustrated eighteenth-century French surveyors because it was in perpetual motion, Europe's cultural borders were constantly on the move, particularly because of the population shifts that accompanied large scale industrialization and urbanization in the nineteenth century. Berghaus's careful dating of his map pointed to one of the core methodological challenges to fixing Europe's cultural groups in territorial space: culturally based borders demanded constant revision to account for rapid demographic changes.

On his map, Berghaus pictured Alsace and Lorraine as contact zones for both "races" (Celts and Germans) and languages (French and German). While the border territories were mostly colored light blue, denoting their Celtic racial character, they were also enveloped within a green language border indicating that the people of the regions were German speakers. Berghaus modeled this French-German language border on a single linguistic study conducted between 1844 and 1847 by Heinrich Nabert.[29] Nabert, whom Berghaus refers to as his "young friend from Braunschweig," was a schoolteacher who performed his language research as a leisure activity during his summer vacations (he also mapped the English-Scottish language border around the same time).[30] But if Berghaus's linguistic and racial borders sometimes relied on a single field study with little verification (the atlas's borders in the Balkans similarly relied on the findings of one field researcher), his atlas nonetheless opened the door to a new way of thinking about the meaning and function of territorial boundaries in Europe.

Despite the failure of German unification in 1848, the proliferation of mapmaking during the 1840s showed that the debate over what constituted a good German national boundary was flourishing well outside the halls of governments. Geographic associations in particular had played an important role in bringing the topic of national boundaries to the attention of the larger German public. As a result, during the 1850s and 1860s, French social scientists came under increased pressure to find a counterargument against German territorial claims to Alsace and German-speaking Lorraine. There was a clear gap between the French-German border that the French mapped at the "natural" border of the Rhine and the French-

VÖLKER-TAFEL.

Indo–Eüropäer:
Kelten ....
1. Eren oder Iren.
2. Galen.
3. Kymren, Walen.
4. _____, Begrade.
Germanen ....
5. Deutsche.
6. Dänen.
7. Schweden } Skandinavier
8. Norweger.
9. Engländer.
Romanische Völker ....
10. Franzosen.
11. Spanier u. Portugiesen.
12. Romanen.
13. Italiäner.
14. Griechen.
15. Walachen.
Slawen ....
16. Serben-Wenden.
17. Tschechen.
18. Polaken.
19. Illyro-Serben.
20. Bulgaren.
21. Russen.
Letten, Littauer.
Albaner.
Osseten, Iron od. Alanen.
Armenier.
Tadschiks, Perser.
Kurden.
Finnen, Uraler.
22. Suomen, Finnländer.
23. Magyaren.
Vaschen.
Semiten.
24. Aramäer.
25. Araber.
Hebräer, über ganz Eüropa
zerstreüt, mit Ausnahme
Norwegens u. Islands; ihre
grösste Dichtigkeit
Berbern.
Kopten.
Turken.
Georgier.
Kaukasier.
Mongolen.
Samojeden.
Jenisseier.

Figure 2.2a "Table of Peoples" from Berghaus's ethnographic map of Europe. Heinrich Berghaus, *Übersicht von Eüropa mit ethnograph. Begränzung der einzelnen Staaten, und den Völker-Sitzen in der Mitte des 19ten Jahrhunderts*, 1846, in Heinrich Berghaus, *Physikalischer Atlas* (Gotha: Justus Perthes, 1848), vol. 2. David Rumsey Map Collection.

Figure 2.2b Cartouche explaining Berghaus's map as an "Overview of Europe, with Ethnic Borders of Individual States, and the Distribution of Peoples in the Middle of the Nineteenth Century, 1846." Heinrich Berghaus, *Übersicht von Eüropa mit ethnograph. Begränzung der einzelnen Staaten, und den Völker-Sitzen in der Mitte des 19ten Jahrhunderts*, 1846, in Heinrich Berghaus, *Physikalischer Atlas* (Gotha: Justus Perthes, 1848), vol. 2. David Rumsey Map Collection.

Figure 2.2c Close-up of the ethnic border zone between French and Germans on Berghaus's map. Heinrich Berghaus, *Übersicht von Eüropa mit ethnograph. Begränzung der einzelnen Staaten, und den Völker-Sitzen in der Mitte des 19ten Jahrhunderts*, 1846, in Heinrich Berghaus, *Physikalischer Atlas* (Gotha: Justus Perthes, 1848), vol. 2. David Rumsey Map Collection.

German border that German atlas makers like Heinrich Berghaus mapped farther to the west, at the "ethnic frontier" of the Vosges Mountains. The Alsace-Lorraine Question that emerged during the 1850s and 1860s can be understood visually as the gap between these two alternative borderlines.

THE ALSACE-LORRAINE QUESTION AND
THE SCIENCE OF NATIONALITY

What happens when two methods for mapping national borders on paper—one based on natural and the other on ethnic boundaries—confront each other in a borderland? A series of direct exchanges between French and German experts in the "science" of nationality help us to reconstruct the meeting of these two perspectives in Alsace-Lorraine. Between 1853 and 1876, European statistical bureaus organized a series of international congresses that provided a forum for debating how to categorize people scientifically into national groups. For many European states, improving methods for ethnic classification was a vital step toward the successful management and pacification of their populations. In the Central and Eastern European states of Prussia, Austria-Hungary, and Russia in particular, the ethnographic question became a key political issue. Evidence from the pan-European congresses reveals the increasing authority of ethnocultural definitions of borders at midcentury, and the diminishing legitimacy of borders found in Europe's physical geography.[31]

For most European statisticians working in the 1850s and 1860s, ethnicity was a powerful indicator of nationality. The Vienna Statistical Congress of 1857, for example, focused specifically on the relationship between ethnographic statistics and the geography of European nationalities. At the meeting, representatives from different European countries exchanged visual techniques for representing ethnic groups using signs, colors, and shades that transformed topographical survey maps into thematic maps.[32] By this time, German cartographers had moved beyond Heinrich Berghaus's mixed racial and linguistic method for defining ethnic borders in his atlas, preferring to focus solely on language indicators as the most simple, rational way for representing ethnic boundaries. A few years later, at the London congress of 1860, German delegates decided to use evidence from thematic maps to publicly confront the French delegation with their ethnic claims to Alsace and German-speaking Lorraine. The German delegates demanded that the head of the French Statistics Administration (*Statistique administrative française*), Alfred Legoyt, explain why the French state refused to conduct a language census in the borderland. From the German delegation's perspective, such hard social scientific evidence would definitively resolve the question of Alsace and Lorraine's "true" nationality. At first, Legoyt countered the Germans' demands by citing the eighteenth-century idea that France's physical geography—its rivers and mountains—decided the proper location of its borders, not cultural factors like language.[33] But then Legoyt shifted his tone. He turned to a second line of reasoning for defending France's claims to Alsace and German-speaking Lorraine: the idea that national borders should not be determined by physical geography or cultural geography, but by a population's "political will."[34]

Charles Schoebel, a founding member of the Parisian-based French Ethnographic Society, was one of the first French scholars to explain the concept of political will in a language that social scientists and statisticians like Alfred Legoyt could understand. Though he was born in Germany, Schoebel married a French woman and spent his adult life in France, where he became an ardent French patriot. With respect to the Alsace-Lorraine Question, Schoebel argued:

> The Alsatians are ethnologically a Germanic people; the language, the physiognomy, the domestic values, and the popular myths and tales from the inhabitants of the Ill [the Alsatian river] . . . demonstrates it above all other considerations. . . . There is therefore, outside of ethnology and linguistics, an ethnic criterion superior to the physical or exterior character of a people; blood, language, and myths are, without doubt, something of great importance . . . but, when it comes to a civilized people, we have, in ethnographic classification, to be very aware of the qualities of the moral being, of [a people's] moral liberty.[35]

In other words, if a researcher were to observe the cultural features of the Alsatian people, referring to "exterior" evidence from language, bodies, and customs, he would logically conclude that they belonged in Germany. But if the researcher looked past these exterior signs of culture that were at the root of ethno*logy* and focused instead on a moral description of the Alsatian people (using the looser, observation-based practices of ethno*graphy*), he would determine that they belonged in France.

In emphasizing political will as the basis for national identity, French social scientists succeeded in exposing the main flaw in the Central and East European approaches to drawing national borders in the mid-nineteenth century. In Central and Eastern Europe, imperial authorities hired experts to decide their population's nationality *for them* based on a "scientific" interpretation of field research, statistical information, and cultural knowledge. These experts simply did not believe in asking people directly: "What is your nationality?" Instead, they preferred to ask indirect questions about language use or, in some cases, religion. A plebiscite, they feared, was a recipe for disaster: it would only confuse ordinary people who could not "properly" identify their nationality. French statisticians and social scientists were thus able to take the moral high ground in arguing that they listened to the democratic will of the people: French victories at the plebiscites in Nice and Savoy in 1861 were a perfect example of the French skillfully using the ballot box, rather than the cultural statistics of nationality, to expand their borders.

Following its annexation to Germany in 1871, Alsace-Lorraine remained a popular case study for experts in the "science of nationality" across Europe, and German maps of Alsace-Lorraine repeatedly emphasized the cultural logic behind their land seizure. One particularly striking map image, *The German Borders against France*, pictured the same cultural boundary lines in Alsace and Lorraine that Berghaus had first established in his atlas during the late 1840s.[36] The image pictured a Prussian eagle, claws extended and tongue protruding in a cry, with patriotic slogans on its wings and a map of eastern France with a language border in its belly (see plate 6). The decision to color the German-occupied areas of the borderland red, moreover, reminded the viewer of the blood sacrificed during the recent war to win back the "authentically" German territory. The image sent a message to all Europeans that the moral justification for a border was found in cultural geography. French scholars, in turn, responded to the German annexation of Alsace-Lorraine by inventing yet another type of unofficial boundary map in the form of a hexagon.

THE HEXAGON: FRANCE'S UNOFFICIAL
NATIONAL BOUNDARY MAP

Just as German nationalists used maps as thinking tools for visualizing the territorial limits of a future German state before unification, French nationalists turned to mapmaking as a means of symbolically reconstructing the wholeness of France after the devastation of the Franco-Prussian War. While France's state-run cartographic bureau accepted the territorial amputation of Alsace and parts of Lorraine in 1871, sending its military surveyors to demarcate the new legal border with Germany in the Vosges Mountains, a separate group of civilian mapmakers began to disseminate alternative images of France that ignored the loss of the border provinces altogether. A number of these civilian mapmakers came from the field of geography, a discipline that acquired newfound respect in French society in the wake of the Franco-Prussian War.[37] One of the most powerful unofficial maps that this new generation of geographers promoted was the image of France as a hexagon. Each point of the hexagon connected to a different French border territory, one of which was Alsace-Lorraine.

The hexagon, which remains a powerful "logomap" for the French nation-state today, is an excellent example of the kind of boundary fantasies that maps promoted in nineteenth-century Europe.[38] Émile Levasseur's *Petit Atlas de la France*, published in the 1870s, was the first to superimpose a hexagonal outline on the relief map of France, and the practice became commonplace in French classroom geography texts throughout the Third Republic. The hexagonal form created a public perception of France's authentic national shape as preordained and almost God given in its mathematical perfection, giving the French a strategy for arguing against the ethnic and linguistic justifications for national borders that other Europeans had embraced. Two of France's leading patriotic experts on geography explained:

> A meridian, which we can regard as an ideal axis, unites the two prominent extremities of the territory in crossing the capital and the center of the figure, dividing France into two halves that are almost symmetrical. From each side of this axis are positioned the sides of a large, irregular hexagon that represents the perimeters of the country whose points are Dunkirk, Brest, Strasbourg, Bayonne, Nice, and Perpignan.[39]

The inclusion of Strasbourg, the Alsatian capital, as one of the six points of the hexagon underscored the political agenda that lay beneath this innocent-looking portrait of France. In popularizing the image of the hexagon as the shorthand visual form of French territory during the turn of the century, French geographers demonstrated that the German annexation of Alsace-Lorraine in 1871 was an illegitimate act that *broke* the hexagon and violated France's right to claim its natural borders. This is why scholars have argued that it was only after 1871 that the hexagon prevailed over the octagon (whose points did not include Strasbourg) as the preferred geometrical image of France (see fig. 2.3).[40] François Schrader, the director of cartography at the French publishing company Hachette, explained the sense of loss that the hexagon expressed: "In its current borders, France is presented vaguely in the form of a hexagon. Before the disasters of 1871, the hexagon extended between the Vosges and the Rhine. We will still conserve this form of France in our thoughts."[41]

While the hexagon gained widespread currency as the unofficial "logo-map" of France during the late nineteenth century, it is important to note that there were also a sizable number of maps circulating in French society that simply ignored the loss of Alsace-Lorraine altogether or left the regions attached but shaded either white, to signify absence, or purple, to signify mourning. Émile Levasseur and Charles Périgot's 1874 classroom atlas pictures Alsace and Lorraine inside France's territorial body but leaves them blank to reveal the loss of their legal status as French regions (see plate 7).[42] Perhaps the most striking example of this "conserved" image of France, with Alsace-Lorraine still attached, was the enormous national map

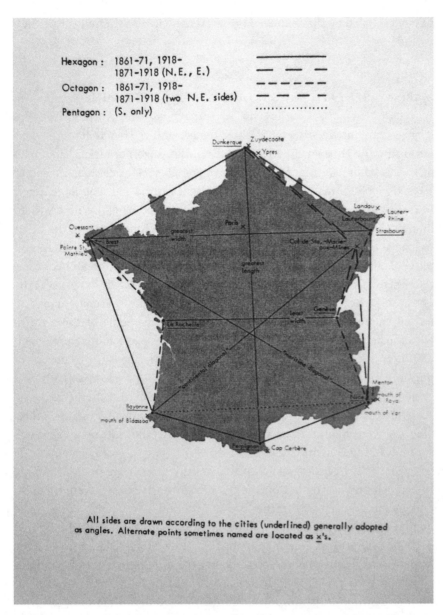

hung behind the center stage at the International Congress of Geographical Sciences held at the Tuileries Palace in 1875 (see plate 8).[43] For the more than 150,000 visitors who attended the Parisian exhibition, paying a modest entrance fee of one franc, the map provided a sort of balm for the painful memories of 1871.[44] In contrast to the reality of territorial amputation, national humiliation, and fragile governance, the France that appeared on the map was strong, unified, and geographically intact. It was a false image of French territory that expressed how French citizens desired to see their boundaries.

MAPPING BENEATH THE SURFACE IN ALSACE-LORRAINE

Rather than justify Alsace-Lorraine's place within French national borders using the abstract shape of an oversized hexagon or an outdated map of France before 1871, another group of French social scientists turned to more tangible forms of evidence buried beneath the soil. Popular interest in France's ancient cultural topography accelerated during the 1860s, when Emperor Napoleon III initiated a project called the Topography of the Gauls to map the roads and military stations that once traversed the Roman Empire in France. It was a historical mapping project that French archeologists referred to as a "topographical restitution" because it sought to resurrect the physical landscape of a lost civilization from the traces of Roman roads and encampments underground.[45] Archeologists located some of these traces with the help of the imperial road service: the construction of a new French road network had inadvertently unearthed the ancient paths of dozens of ancient Roman routes. In Alsace, research for the Topography of the Gauls project was a joint enterprise between Alsatian historical preservation associations and state authorities, reflecting a growing interest in France, among government officials and citizens alike, in the cultural and historical roots of the French nation.

After 1871, historical maps of Alsace only grew more popular among the French public. The notion that France's historic borders were timeless, and lay buried deep within the layers of the soil, gave French citizens the mental image of a stable and enduring nation-state, even after the trauma of defeat.[46] One of the foremost supporters of historic research into France's ancient topography was the conservative writer Maurice Barrès, who encouraged his fellow French citizens to embrace a "cult of the dead."[47] In Alsace, Francophile anthropologists incorporated Barrès's fascination with the French people's dead ancestors into their scientific research, comparing the skeletons buried beneath Alsatian soil with those buried in the rest of France. Meditating on the question "To which race do the Alsatians belong?" Ferdinand Dollinger, an expert on Alsatian anthropology wrote:

> The evolution of an innumerable line of ancestors determines our organs, our senses, our thoughts, our acts: in sum, our temperament and our personality. . . . It is not enough to analyze our body and our soul; we must turn back to the past, the farthest back that we can sense, to discern the essential traits of those who formed the mold of our physical and mental physiognomy: we must define and study our race.[48]

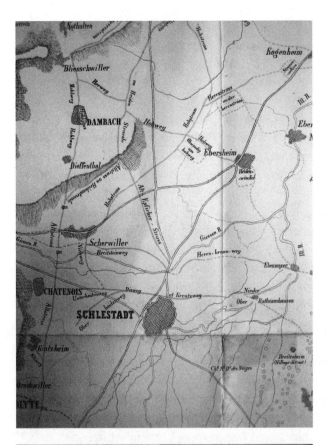

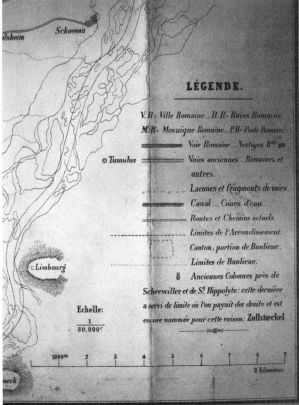

Figure 2.4a Napoleon III's Topography of the Gauls project aimed to map the history beneath the ground in Alsace, bringing back to life the network of ancient Roman roads that had once traversed the border region. Charles-Gabriel de Morlet, *Topographie des Gaules: Notice sur les voies romaines du Département du Bas-Rhin* (Strasbourg: Imprimerie de Veuve Berger-Levrault, 1861). Photo. et coll. BNU Strasbourg.

Figure 2.4b Map legend indicating the different types of Roman roads in Alsace. Charles-Gabriel de Morlet, *Topographie des Gaules: Notice sur les voies romaines du Département du Bas-Rhin* (Strasbourg: Imprimerie de Veuve Berger-Levrault, 1861). Photo. et coll. BNU Strasbourg.

One of the most striking examples of the anthropological attempts to classify the people of Alsace as "racially" French was a series of excavations conducted by Doctor Edmond Blind at the University of Strasbourg.[49] The doctor carried out an extensive study of seven hundred skulls removed from medieval crypts and ossuaries located in different spots in Alsace: the villages of Dambach, Saverne, Lupstein, Scharrachbergheim, Kaysersberg, and Ammerschwihr. Blind was clear about his larger research ambitions: he explicitly stated that he wanted to map out the "ethnological frontier" between modern France and Germany. His anatomical observations and bone studies resulted in the following classification of Alsatian skulls: 84.56% Brachycephalic, 13.71% Mesocephalic, 1.7% Dolichocephalic.[50] Because most modern Frenchmen had the Brachycephalic type of skull while Germans had the Mesocephalic type, Blind linked the Alsatians unequivocally to French ancestry.[51] The study had important implications. In driving a wedge between physical ancestry and acquired culture (language, customs, etc.), the Francophile anthropologists were able to rival the German cultural claims to Alsace-Lorraine. In their view, they had succeeded in establishing an authentic French-German border based on race than ran *deeper* than a more superficial kind of cultural border based on ethnicity or language.

Meanwhile, German archeologists conducted research to map the historical layers beneath Alsace and Lorraine according to their own priorities, sometimes calling in garrisoned troops to help them with excavations.[52] The *Historical Map of the German Empire*, a collective research initiative of private German archaeological and historical associations begun in 1891, was a feat both of technological achievement and of patriotic determination.[53] In creating map images of the German Empire at different points in its history (well before it was formally unified in 1871), the archaeologists and historians hoped to rid German historical studies of the narrow-minded particularism of regional and local studies and create instead a big-picture, German-national perspective on history. Figure 2.5 indicates some of the major towns and landmarks in the French-German border area that were part of these larger German networks.

According to Friedrich von Thudichum, the professor from the University of Tübingen who organized the historical cartography initiative, the goal of the maps was to "bring to view . . . collective and important historical relationships" between German lands.[54] The thematic maps highlighted historical information that included religious relationships, networks of German universities and gymnasiums, royal residences, coin production, imperial cities, trade routes in the Middle Ages, toll stages on the German rivers, the history of the German customs union, the history of German railroads and canals, the court system, pilgrimage routes, unions of princes and nobles, mining, the book trade, and wine production.[55] Making the

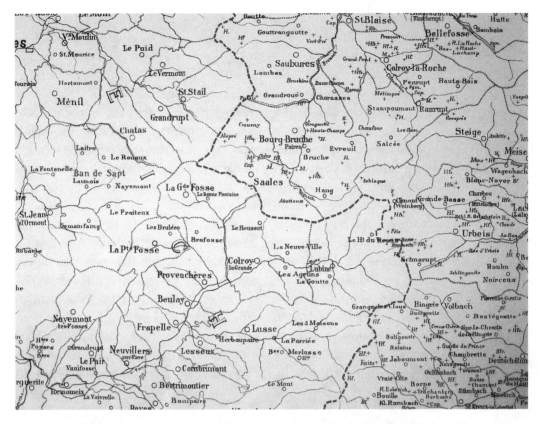

Figure 2.5 Historical map of routes and settlements in Alsace. The map was part of a collective project undertaken by German historical associations called the *Historical Map of the German Empire*. Dr. Friedrich von Thudichum, *Historisch-Statistiche Grundkarten: Denkschrift* (Tübingen: Verlag der H. Laupp'schen Buchhandlung, 1892). Photo. et coll. BNU Strasbourg.

maps available to archives, schools, clubs, gymnasiums, and military academies, the association hoped that the visual representations of Germany's historical geography would "significantly improve the feeling of togetherness and German interests for the people from the farthest German border provinces."[56] The association's goal, in other words, was to create a synthesized view of the German Empire as a single, unified cultural space, whose borders were rooted in a long-standing history of collective social relationships.

Germany's flourishing archaeological projects in Alsace-Lorraine would end with one last triumphal dig: an unexpected discovery in a World War I trench. In 1916, German soldiers at Norroy-lès-Pont-à-Mousson, on the left bank of the Mosel River near Metz, came across a beautiful altar to the god Hercules Saxsetanus, dating from AD 80. The head of the Metz archaeological museum, J. B. Keune, personally traveled to the war front to photograph and arrange for the transportation of the altar. Keune's actions were supported by a special protocol that the German Army had established earlier in the war to protect any objects of cultural heritage discovered along the battlefront (*Schutzverwahrung von Kunst und Kulturwerken*). The altar's transportation was made easier by the fact that a French mine had exploded it into multiple pieces. There is still a photograph of the museum director, standing fully underground, inside the trench, accompanied by German soldiers in waistcoats with binoculars in their hands.[57]

Meanwhile, as German archaeologists and soldiers dug into the soil of Alsace-Lorraine, France's wartime government quietly called together a special committee of social scientific experts to study the potential location of a postwar French-German boundary. The list of committee members included some of the leading lights of the French academy, including geographers Paul Vidal de la Blache, Emmanuel de Martonne, and Lucien Gallois, and historians Ernest Lavisse and Christian Pfister. In 1918, the wartime committee produced an atlas, *Alsace-Lorraine and the North-East Frontier*, which included a series of thematic maps that each illustrated the border area's long-standing historic, cultural, and economic ties to France.[58] The decision to publish the committee's research findings in the form of an atlas revealed the extent to which the French had come to embrace territorially based nationalism by the early twentieth century. When it came time to justifying Alsace-Lorraine's return to France after 1918, French officials were prepared not only with the argument of "political will," but also with cartographic evidence.

THE POWER OF NATIONAL BOUNDARY
MAPS IN INTERWAR EUROPE

The utopian concept of a "good border," which European nationalists had invented with the help of maps during the nineteenth century, became tremendously influential after World War I, when Europeans sought to draw new national borders that would ensure a lasting peace. At the meetings that resulted in the Treaty of Versailles in 1919, representatives of Central and Eastern European countries relied on the concept of cultural boundaries—both linguistic and ethnic—to negotiate the legal borders of their new nation-states.[59] There were some serious obstacles, however, in using cultural boundaries to guide the demarcation of legal state borders. In many areas of Europe, ethnic and linguistic populations were too mixed up to be divided by any meaningful cultural boundary. The solution, several European countries decided, was to use state borders as filters to "sort out" ethnic populations from one another. Massive population transfers, such as the 1923 population swap between Greece and Turkey, ensued with disastrous results.[60] But the notion that state boundaries should align with the limits of ethnically defined national populations was not restricted to the countries of Central and Eastern Europe. France, which scholars have often viewed as Europe's exception to the ethnocultural nation, participated fully in the ethnically defined interwar border system.[61] Beginning in 1918, Alsace-Lorraine became the site of a French experiment in transforming its political border with Germany into an ethnically pure one.

Shortly after French soldiers liberated Alsace and Lorraine from German control, French "triage" commissions assigned each resident of the border area an ethnic identity card in order to distinguish "real" Alsatians and Lorrainers from the thousands of "ethnic" Germans who had emigrated after 1871 to serve as teachers, judges, professors, bureaucrats, architects, railroad engineers, and priests.[62] The French government issued "A" cards to those inhabitants with two Alsatian or French parents, "B" cards to those persons with one Alsatian parent, "C" cards to residents of foreign (non-German) origin, and "D" cards to residents of German origin, even if they had been born in Alsace-Lorraine. As a result of the measures, lifelong residents of Alsace-Lorraine who were determined to have German ancestry were forced to leave the borderland, confirming that French national identity in postwar Alsace-Lorraine "was determined by blood."[63] In total, an estimated 110,000 residents from Alsace, and nearly 100,000 from Lorraine, emigrated from the border area after World War I because they lacked proper national identity cards, or out of fear of retribution.[64] Photographs from 1919 and 1920 show thousands of men, women, and children crossing the French-German border on foot across the Rhine Railway Bridge, carrying whatever possessions they could in their hands.[65] Some would later write to the French government and request special visas to retrieve their libraries and furniture.[66] Many would write memoirs that nostalgically recalled their deep emotional and personal attachments to Alsace-Lorraine.[67] According to Strasbourg's statistical office, the postwar purge of residents with German ancestry had such a dramatic impact that it caused the population of the city to drop back to its 1905 levels.[68]

The compulsory population transfer of ethnic Germans out of Alsace-Lorraine had a direct bearing on German irredentist politics during the 1920s and 1930s. A group of professors who had been purged from the University of Strasbourg, for example, founded their own institute in Frankfurt, the Wissenschaftliche Institut der Elsass-Lothringer im Reich, which devoted itself solely to the study of Alsace-Lorraine. The institute published a journal, the *Elsass-Lothringische Jahrbuch*, which continued the social scientific research on Alsace-Lorraine that had absorbed German scholars during the turn of the century. In 1931, Professors Werner Gley and Georg Wolfram published an atlas that reiterated the full range of German cultural claims to Alsace-Lorraine through the visual narrative of maps.[69] According to Gley and Wolfram, the objective of the *Elsass-Lothringer Atlas* was to illustrate the political, commercial, and cultural development of Alsace-Lorraine since prehistoric times from a "scientific and unprejudiced view."

Created amid an atmosphere of increasing tensions between France and Germany, Wolfram and Gley's atlas, together with other cartographic publications, demonstrated how maps acted as a conduit for territorial re-vanchism in interwar Europe.[70] The maps in the *Elsass-Lothringer Atlas* each presented a different set of facts about the German identity of Alsace-Lorraine, ranging from place-name terminations, to architectural distributions, to historical networks of coin exchange. The circulation of the atlas among German scholars and elites meant that Germany had not yet laid the dispute over Alsace-Lorraine to rest; the maps kept German alternatives to the post-Versailles border alive.

Meanwhile, once the purge of "ethnic" Germans from Alsace-Lorraine was complete, a progressive group of French historians proposed their own alternatives to the ethnic border logic that dominated the interwar years. During the 1920s and 1930s, the University of Strasbourg was home to the Annales school, a contingent of scholars who preferred to study the history of deep, slow-moving shifts in social structures and environments to the "superficial" history of events. One of the most prominent members of the Annales school, Lucien Febvre, argued that it was important to take a long-term view of the Rhine River's legitimacy as a national border. Febvre believed that the Rhine was not a natural border at all, but rather an "international river" that over the centuries had served as a long-standing point of connection between the French and German peoples. In 1931, the Alsatian Bankers Association, a civil association that had an economic interest in promoting this fresh interpretation of the Rhine boundary, commissioned Febvre to lay out his alternative philosophy of borders in a published volume.[71] "The Rhine," he wrote, "remains a river that joins together, in spite of political hatred and conflicts."[72] It was, he insisted, "a mediator between West and East."[73] While World War II swiftly brought Febvre's cooperative vision of the French-German border to an end, his ideas did serve as inspiration for the new kind of French-German boundary that would come to fruition in the second half of the twentieth century.

During the nineteenth and early twentieth centuries, a new genre of boundary map helped Europeans to visualize their land in terms of ethnic, historical, and even racial topographies. While scientific survey maps represented the sovereign territory of states, these thematic maps framed European geography as a mosaic of national homelands. The geographers, linguists, ethnographers, archaeologists, anthropologists, and historians who created these maps overlooked the official border that states established in Alsace-Lorraine through legal treaties in favor of more "natural" borders that could serve as clean lines of separation between French and German populations.

In doing so, they created compelling arguments for territorial revanchism and kept alternative locations for the French-German border alive in public debate. Like government-sponsored survey maps, thematic maps of Alsace-Lorraine tended to create a feeling of distance between the map reader and the land that appeared on paper. They were based on "scientific principles" of nationality rather than on the emotional or sensory experience of land. Nonetheless, thematic maps of national boundaries help us to uncover the salient principles that French and German elites used to define the limits of their national communities during the nineteenth and twentieth centuries. They marked an important step toward "producing" national territory in the same way that scientific survey instruments "produced" states. The following chapter explores a particular form of thematic mapmaking that became especially politicized in the conflict over the French-German borderland: language mapping.

CHAPTER 3

Language Maps

In Strasbourg today, street signs appear in both French and German, and the cafés and *Weinstuben* in the Alsatian countryside bear names with distinctive regional charm. But different national and regional languages did not always find tolerance in Alsace-Lorraine, where words were once hotly contested cultural symbols. During the rise of European national movements in the nineteenth century, the perceived relationship between language and cultural identity formed the basis for powerful nationalist narratives. Language maps, a visual medium first invented during the early nineteenth century, helped Europeans to project their linguistically defined nations across territorial space. The maps stored linguistic knowledge in visual form, serving as depositories for language data gathered from surveys, censuses, and field observations. Politically, maps of national language zones created symbolic ruptures between European peoples, suggesting to viewers a correlation between linguistic divides and national borders. By the early twentieth century, language borders had achieved meaningful currency in European political culture, shaping popular conceptions of national space and influencing the placement of Europe's new national borders in the Treaty of Versailles.

Alsace and Lorraine's historically strong connections to the German-speaking world, even in the centuries following their annexation to France, made them the subject of extensive experimentation with language mapping during the age of nationalism. While few among the borderland's population spoke formal German, or *Hochdeutsch*, many of the ordinary people living in Alsace spoke a German-based dialect called Alsatian, while in eastern Lorraine they spoke a dialect called Franconian, also with German roots.[1] For nineteenth-century German nationalists, freeing this sizable group of German speakers "trapped" inside French borders became a serious public cause. But in order to prove their linguistic claims to Alsace and German-speaking Lorraine, German nationalists first needed to identify the precise geographic limits of their language borders. This created an intriguing methodological problem. How exactly does one uncover the point in territorial space where the French language "ends" and the German language "begins"? In order to locate, survey, and render visible the elusive French-German language border, nationalists needed a new set of mapping techniques.

When Europeans drew their language borders, they did so according to vastly different criteria from their state borders. Language "surveyors" used an array of evidence, including travel narratives, historical documents, state-sponsored censuses, and on-site interviews, to map clean borderlines between linguistic populations. It was a great challenge, however, to establish fixed language borders that promised not to move. Europe's language zones were inherently malleable: they changed with the rise of new commercial and communications networks, particularly those brought by the Industrial Revolution and the development of railroads. Towns located on the frontiers of European borderlands—including Bohemia, Tyrol, Posen, the Cerdanya, and Alsace-Lorraine—shifted over time from one language to another, depending on economic and political opportunities. The language borders that appeared on thematic maps of modern Europe must therefore be interpreted as imperial arguments rather than as transparent reflections of reality.[2]

As pictorial arguments about the cultural identity of Alsatians and Lorrainers, language maps revealed specific differences between French and German visions of their national space. For example, Germans developed an entire vocabulary to describe the territorialization of language on European land, while the French did not.[3] There existed no French equivalent of the German concept of a *Sprachgebiet*, or "language area," a term frequently employed in nineteenth-century German political texts. The visual construction of the German *Sprachgebiet*—through German nationalists' merging of linguistic knowledge with territorial maps—provided a blueprint for a European order organized according to language families. In the decades before and after the annexation of Alsace-Lorraine in 1871, German nationalists accumulated language evidence from the borderland through numerous language inquiries, or *Sprachproben*. Using the information gathered through these surveys, mapmakers established divides—alternately termed *Sprachwege* (language paths), *Sprachstriche* (language lines) or *Sprachgrenze* (language borders)—to create a neat visual border between French speakers and German speakers. These language maps, mass produced thanks to advances in printing technology, drew the attention of the elite classes in Berlin and other German cities, who sought tangible cultural justification for

Germany's imperial venture in the French-German borderland. With the exception of an intriguing and innovative attempt to map France's multi-lingual empire under Napoleon Bonaparte, French mapmakers displayed little interest in cartographic expressions of language. It is precisely from such "cartographic silences," however, that we can draw important conclusions about the shifting role of language in French discourses on national identity.[4]

But just as language maps provide excellent historical sources for comparing French and German philosophies of nationality, they also offer tremendous insight into the invention of regional identities in nineteenth-century Europe. Alsatian language researchers were generally skeptical of the binary concept of a "French-German" language border, focusing instead on dialect research that affirmed a uniquely Alsatian territorial space. During the turn of the century, Alsatians drew and popularized never-before-seen language borders within their own region. By doing so, they demonstrated the power of "counter-mapping" as a form of cultural resistance to the dominant nationalist agendas in the borderland.[5] In comparing what motivated individuals to create their language maps of Alsace-Lorraine—and whether states decided to support their efforts—we learn how cartographic practices came to play a decisive role in shaping the debate over the borderland's disputed cultural identity.

FROM VISIBLE TO INVISIBLE: FRANCE DRAWS AND ERASES ITS LANGUAGE BORDERS, 1789–1870

For centuries, French monarchs displayed an extraordinary degree of tolerance for the variety of languages spoken on French soil, making only weak attempts to impose French as an official state language.[6] It was not until revolutionaries upended the Old Regime in 1789 that France's linguistic diversity began to be seen as an impediment to national unity.[7] In Alsace, zealous Jacobin revolutionaries, including one of the chief architects of the Terror, Antoine Saint-Just, favored a series of heavy-handed measures to purge the German language from the region and to align more closely France's political and linguistic borders. Although short lived, the so-called linguistic Terror[8] revealed one of the first moments in modern French political culture when the French government turned to language, a key component of ethnically defined nationality, as a means to unite its people. In Alsace, as well as in Brittany, Corsica, and parts of Languedoc, French administrators planned a series of language schools to "Frenchify" the state's border populations. By 1794, at the height of the linguistic Terror, the municipality of Strasbourg had purged all visible symbols of German from public view, including signs from stores, workshops, and boutiques.[9]

Local reactions to the revolutionary government's language initiatives in Alsace were mixed. Map toponyms provide some clues to the complexity of revolutionary language politics in the border area. During the 1790s, some towns in Alsace Frenchified their names (Bockenheim became Saar-Union, and Kaysersberg became Mont-Libre), while other towns "Germanized" their names (Saint-Louis became Heysersberg, and Saint-Louis-lès-Bitche became Münzthal).[10] The French language, therefore, did not have the same clear-cut meaning in the revolutionary atmosphere for Alsatians as it did for the Parisian revolutionaries. While some Alsatians saw German as backward and embraced French as the language of national pride and political progress, other Alsatians who had struggled under the rule of the French monarchy associated the French language with despotism and the unequal class relationships of the Old Regime. As the favored language of "the people," German was, in the minds of these Alsatians, a language that fit closer to the Revolution's ideals of popular empowerment. Alsatian locals and Jacobins therefore disagreed over the idealization of French as the language of the nation.

In the wake of the revolutionary violence, Napoleon Bonaparte continued to monitor the linguistic makeup of France's border areas, but he took a far less hardline stance than the Jacobins. Building an empire rather than a strong centralized republic, Napoleon sought linguistic information that would help him to manage his expansive multicultural territory. As a native speaker of Corsican, moreover, he had little personal desire to crush regional languages.[11] In 1806, Napoleon commissioned Charles-Étienne Coquebert de Montbret, head of the French Empire's recently founded Bureau of Statistics, to map the boundaries of what he considered to be France's five "mother languages": French, German, Flemish, Breton, and Basque.[12] It was in many ways an innovative and pathbreaking project that generated new ideas about the meaning and structure of European borders. Mapping cultural boundaries, Coquebert de Montbret learned, required inventive cartographic methods capable of gathering, analyzing, and representing social information. Completed in 1806, his language maps—though lacking the sophistication and broad public audience of later works of linguistic cartography—signaled the stirrings of a powerful nineteenth-century desire to see European land through the framework of cultural geography.

To research the location of France's language boundaries, Coquebert de Montbret had a vast government infrastructure at his disposal.[13] But what kind of linguistic information was needed and from whom should he gather the information? How exactly does one locate the geographic meeting points of languages? The scientific tools and astronomical observations that the French state relied upon to map its official, sovereign borders were clearly of little use for mapping culturally based boundaries. Coquebert de Montbret therefore had to improvise. Rather than dispatch teams of Parisian-based surveyors out to the French countryside, he decided to ask local notables and government prefects from across France to send him reports on the languages spoken in their geographic areas. The information that he received in return was a messy patchwork of locally generated linguistic knowledge, including historical documents, embellished travel accounts of encounters with locals who spoke bizarre tongues, samples of poetry and folk songs, and even translations, in the local patois, of the parable of the prodigal son.[14] Not only was the type of document that he received problematic; there was no consistency in linguistic terminology. One prefect referred to what is now called the Alsatian dialect as everything from a "dialect" to an "idiom" to a "German language."[15] All the people involved in the survey, it seemed, perceived the languages around them, and the cultural identities that they represented, differently.

Despite the challenges of working with this unwieldy range of social evidence, Coquebert de Montbret synthesized the information as best he could, hand painting a series of language borders on administrative maps of France's recently created departments (see plate 9). Crude as they were, Coquebert de Montbret's maps were highly innovative. Linguistic geography, relegated in the eighteenth century to the travel accounts of bourgeois gentlemen, was now visible from a commanding "god's-eye view."[16] Images of language borders, just like state borders, could now be printed on paper maps and distributed, modified, and exchanged. For Napoleon's imperial administrators, in the Bureau of Statistics and elsewhere, the maps offered a quick and easily discernable overview of the cultural background of the populations under their control.[17]

Coquebert de Montbret's language maps were, however, still extremely limited in their ability to reach a broad public audience and in their capacity to tackle questions of national identity. There was no connection whatsoever between the French-German language border pictured in plate 9 and the idea of a "national border" between France and Germany. When Coquebert de Montbret labeled half of the Department of the Meurthe *communes allemandes*, or "German communities," he did not mean that he thought of the people living there as German nationals; they were simply French people who spoke German. This tolerance for non-French languages would soon disappear from later nineteenth-century French regimes. Feeling pressure from Central and Eastern European countries, where national activists were turning to linguistic geography as their primary basis for staking territorial claims, subsequent French governments chose to hide their country's internal language borders from view. For the remainder of the nineteenth century, language borders would be strikingly absent from the French cartographic archive, a powerful example of "cartographic silence." In German-speaking Europe, on the other hand, language mapping flourished as the nineteenth century progressed. Not only did Germans develop new methods for language surveying; they also transformed language maps into familiar objects for all to see. They became public communications devices capable of making powerful border claims.

VISIONS FOR A FUTURE GERMAN EMPIRE

The increasingly serious dilemma that Alsace and Lorraine's linguistic makeup posed for French nationalists in the nineteenth century contrasted sharply with the strong hand that it dealt the Germans in claiming the borderlands as part of their national space. German nationalists maintained that language was something that marked people from birth, fixing their innate cultural identity from the time that they entered the world. Romantic-era thinkers such as Herder and Fichte celebrated language as the essence of German-ness, while Napoleonic-era nationalists such as E. M. Arndt called language borders "the only worthy natural borders" and proclaimed that the German fatherland "extends as far as the German language is heard."[18] Scholars such as the Brothers Grimm, meanwhile, traveled across German-speaking Europe in the early nineteenth century, collecting information on German folklore and language for an eager public audience.[19] It was only in the years preceding the revolution of 1848, however, that Germans first attempted to map the German *Sprachgebiet* from an aerial view.[20] During this period of great hope for German national unification, language maps became idealized spaces for nationalists to visualize how far the future borders of the German state should extend. Unlike official state borders, however, the language borders (*Sprachgrenze*) that German nationalists invented during the 1840s did not require any negotiations or compromises with neighboring states. They were creative tools of the geographic imagination, helping Germans to imagine an ideal territorial future into being.[21]

In 1844, Karl Bernhardi, a future deputy to the Frankfurt Parliament of 1848, published his *Language Map of Germany*, the first map to picture the lay of Germany's linguistic landscape. While the French could rely on their Bureau of Statistics to organize their first language map survey during Napoleon's reign, German nationalists like Bernhardi had no such centralized state bureaucracy to help them gather linguistic information. Instead, Bernhardi had to rely entirely on information solicited and organized by members of German civil society. Over the span of nine years, thirteen private historical associations sent Bernhardi language information from a range of German-speaking states, including travel writings, historical articles, opinions of local elites, and language census numbers.[22] Like the French cartographer Coquebert de Montbret, Bernhardi never visited all

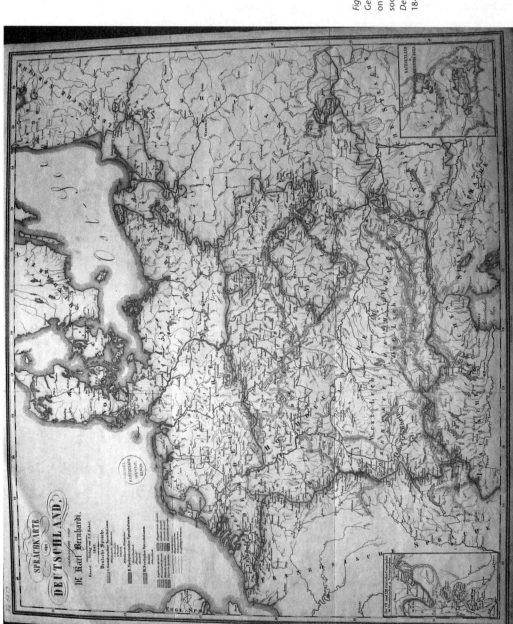

Figure 3.1 Karl Bernhardi's *Language Map of Germany* was the result of nine years of work on behalf of thirteen German historical associations. Karl Bernhardi, *Sprachkarte von Deutschland* (Kassel: Verlag von J. J. Bohné, 1844). Photo. et coll. BNU Strasbourg.

the linguistic borderlands that he mapped; he was a "compiler," an armchair mapmaker who synthesized language evidence sent from afar into the visual form of a map.

Bernhardi based the path of his map's French-German language border, indicated with a green line (see detail in plate 10), on the grassroots-generated language surveys of Heinrich Nabert, who had spent his summer vacations hiking through Alsace and Lorraine to interview locals about their language use.[23] Bernhardi's *Language Map of Germany* thus reveals the astounding capacity for a European with no professional training in cartography to invent and publicize his own territorial border in the nineteenth century. It was not a perfect border, Bernhardi acknowledged. If we look closely at the map, it is evident that he decided to blur the French-German language border ever so slightly, a subtle acknowledgment that the precise location of language boundaries could never be determined with absolute certainty.

In spite of its lack of precise borderlines, however, Bernhardi's *Language Map of Germany* established a powerful relationship between the German *Sprachgebiet* and the idea of German national land. In symbolically uniting the three major German-speaking areas of Europe—High German, Low German, and Nordic—the map made a compelling visual argument for the collective identity of the different language groups. Moreover, the map argued that the collective cultural identity of the German people should overpower the legal state boundaries that continued to divide Germans from one another. Official state boundaries, in fact, were nowhere to be found on the map. Bernhardi had deliberately removed them, replacing them with language borders that enclosed a unified German national space.

To gauge the impact of Bernhardi's language map on German public opinion in the 1840s, we must expand our analysis beyond the map's border semiotics and explore how the paper image circulated in German society. From the start, Bernhardi had in mind a large audience for his map. Cheaply printed on a map sheet of Europe copied from Adolf Stieler's popular *Hand-Atlas*, Bernhardi distributed his map to dozens of German civil associations and middle-class politicians. At a time when German nationalists were thinking about how to overcome the immense challenges of unification, Bernhardi's simple image of the German *Sprachgebiet* offered a much-needed vision of a collective cultural geography.[24] His map demonstrated one of the chief reasons why language borders became so influential in nineteenth-century Europe: they were practical thinking tools for nations "in the making."

While the failure of the Frankfurt Parliament dashed the hopes of unifying German territory in 1848, language maps continued to serve as powerful communications devices for German nationalists. Beginning in the 1850s, German statisticians joined other Europeans in a series of international congresses that revolved around the use of language statistics for categorizing populations according to nationality.[25] The statisticians hoped that government-administered language censuses could improve the sporadic nature of the linguistic information that mapmakers had been processing with some difficulty since the beginning of the century. German nationalists, meanwhile, believed that establishing partnerships with state statistical bureaus—a form of public-private collaboration—could help them build a strong "scientific" base of evidence for German unity. One prominent German statistician, Richard Böckh, combined the linguistic underpinnings of German Romantic nationalism with the modern tools of empirical science.[26] Rather than rely on the haphazard linguistic evidence of the past—travel narratives, historical documents, or subjective observations—Böckh turned to the population census, a powerful modern state tool for managing populations, to construct elaborate tables of numerical evidence to demonstrate the vast expanse of German speakers across European territory.[27] His positivist faith in the capacity for statistical analysis to delineate German language boundaries objectively, however, would be tested by a series of methodological dilemmas.

Böckh's language research rekindled a long-standing debate among nineteenth-century European language statisticians over the value of surveying a population's "mother tongue" (*Muttersprache*) versus its "language of everyday use" (*Umgangssprache*). Summarizing the views of his German colleagues, Böckh argued that the mother tongue was the only sound criterion for determining the nationality of a population. To measure the language of everyday use was to take into account foreign influences on a people's authentic national roots and distort the location of Europe's "organic" national borders. Böckh's attachment to the mother tongue as a statistical category reflected a conception of national identity that bordered on the biological: "Language is a symbol of ancestry [*Abstammung*]; it also has a physical basis; it is passed down through the bodily nature [*körperliche Natur*] of man and finds itself dependent on the same to a large degree; [language] has definite organic foundations, which are differentiated according to the ancestry of men, and the individual, at his birth, already appears physically predestined for a certain language."[28] Not all European statisticians agreed with Böckh's philosophy. A notable counterexample to Böckh's research method was that of Baron Karl von Czoernig, director of the Austrian statistics administration and the author of the *Ethnographic Map of the Austrian Monarchy*, completed in 1855. In his attempt to create a territorial image that minimized cultural differences between the Austro-Hungarian Empire's various nationalities, von Czoernig preferred the category of the language of everyday use to underscore the empire's capacity to assimilate its people through a shared imperial language.[29]

Böckh's statistical study of Europe's German-speaking lands would set an important precedent for late nineteenth- and twentieth-century German state policies and private initiatives regarding minority German populations in other European states. According to his argument, any state that attempted to change its population's "predestined" mother tongue violated the principle of nationalities and was guilty of "imperializing" behavior.[30] Acting on his personal beliefs, Böckh became a founding member of the German School Association for the Preservation of Germans in Foreign Countries, an organization that supported German schools and culture among the German diaspora in areas such as Tyrol and Bohemia.[31] Böckh felt the same sympathy for the German-speaking population in France that he felt for German-speakers in these other parts of Europe and provided his readers with a statistical table of the German-speaking population in France using data mined from historical documents dating to the Holy Roman Empire.[32] In 1872, when the German government conducted its first language census in Alsace-Lorraine, the results reiterated Böckh's "empirical proof" of German rights to the borderland: the census indicated that the overwhelming majority of the population in the annexed territories spoke German.[33]

LANGUAGE AS SYMBOLIC CONQUEST: RECOVERING ALSACE-LORRAINE'S "AUTHENTIC" PLACE-NAMES

Alsace-Lorraine's transition from French to German rule, beginning with the German Army's occupation of the region in the fall of 1870 and formalized with the Treaty of Frankfurt in 1871, was strongly influenced by the perceived connection between the borderland's linguistic geography and its nationality. Historians of cartography have long pointed to toponyms as important imperial signifiers on a map: naming a place can be interpreted as an act of symbolic appropriation and a step toward imposing cultural hegemony on a foreign land.[34] Even before the dust had settled on the battlefields of the Franco-Prussian War, the German imperial government began to return many of Alsace-Lorraine's French-sounding place-names to their "original" German forms, many of which were resurrected from the time of the Holy Roman Empire. To assist in this first act of symbolic territorial possession, the German administration turned to Richard Böckh, who was temporarily employed in Strasbourg's statistics administration from 1870 to 1872, and Heinrich Kiepert, the renowned professor of geography in Berlin and the former director of the Geographical Institute in Weimar. Primarily an expert in mapping foreign lands, Kiepert directed his cartographic expertise toward Germany during its period of unification. Together, Kiepert and Böckh would complete the German Empire's utopian vision of Alsace-Lorraine as a long-lost German land; they would turn back the clock and inscribe the region's imperial German past onto its nineteenth-century landscape.[35] With their help, the German imperial state handed out new names to Alsace-Lorraine's towns and villages, an authoritarian measure that involved no democratic debate.

To gather their language evidence, the two experts from Berlin, in collaboration with regional archivists, pored over old ledgers and documents in search of the borderland's historic place-names. Their research frequently involved undoing the name changes that the French monarchy had made when it conquered the border region in the seventeenth and eighteenth centuries: the French had replaced the sounds "u" with "ou" (Strassburg became Strasbourg); "weiler" with "weier"; "willer" with "wihr"; and "dorf" with "troff."[36] Theoretically directed by historical research and linguistic analysis, the search for an Alsatian or Lorrainer town's "authentic" German name was unmistakably haphazard. Lists of old and new place-names were littered with question marks and multiple suggestions.[37] Upon government request, local authorities scurried into dusty town halls for documents that would unlock the mystery of their town's historic name. In some cases, the German imperial administration did allow towns with historic French toponyms in the western part of Alsace-Lorraine to keep their place-names as long as the majority of the population spoke French. One mayor, for example, was able to verify that the authentic name of his French-speaking Lorrainer town was French by uncovering a 250-year-old church document in a parish building.[38] Amid the widespread linguistic Germanization efforts, the lingering question of what to do with these three hundred or so French-speaking towns would shape the German Empire's future language mapping projects in the border region.

THE MUTABILITY OF LANGUAGE BOUNDARIES

In the years leading up to unification, German nationalists had idealized language borders as the only "true" borders of the German nation. But when it came to drawing a new legal border with France in 1871, the newly unified German state ruled out the idea of using a language boundary as its guideline. Instead, Prussian military officials convinced the German chancellor Otto von Bismarck to negotiate a boundary with France that satisfied Germany's need to defend itself in a future war. As a result, the German state created in 1871 significantly overreached its western language borders: three hundred thousand French speakers, mostly from Lorraine, became part of the German Empire.[39] The year 1871 therefore marked a significant turning point in the function of German language maps. After 1871, mapping language borders ceased to be an act of idealism on the part of German nationalists, and instead became a means for guiding the empire's assimilation of its culturally mixed border areas.[40] German nationalists no longer viewed language borders as fixed entities; instead, they became believers in their mutability.

In monitoring changes to the language border within Alsace-Lorraine, as they did in the Polish-speaking regions in eastern Prussia, Germans could mark the progress of their linguistic assimilation policies. The first language survey conducted after 1871 was Heinrich Kiepert's *Language Map of Alsace-Lorraine* (see plate 11 and plate 12). A curious combination of contemporary and historical perspectives, the map included population data from the recent 1872 census, on-the-ground linguistic observations from Kiepert's summer research trips, and evidence from the historic French-German border. His map transformed each of the districts in Alsace-Lorraine into a distinct *Sprachgebiet*, organizing them into the following categories: "French since antiquity," "now French, partially or entirely German in the seventeenth or eighteenth century," "mostly French," "German and French, almost equal parts," "mostly German," and "German."[41]

Kiepert's methodological description of his language survey revealed the contradictions between his belief in the permanence of historic language borders and witnessing with his own eyes the rapid shifts in the French-German language border within a matter of decades. One of the advantages to observing the language border on foot was that Kiepert could meet with locals and establish a genealogy of spoken languages in the border area. For example, Kiepert described a meeting in a forest with a farmer from the village of Waldersbach who spoke French, but who told him that his father had spoken French and German, and his grandfather had spoken only German. Trusting the linguistic heritage of the farmer's ancestors over his current language of use (reminiscent of Böckh's defense of the "mother tongue" concept), Kiepert concluded that the farmer's French-speaking village was in fact a "purely German colony."[42] Kiepert also cited his interviews with German railroad engineers, who noted the heavy influx of German-speaking Alsatian miners into historically French-speaking communities, as evidence that the modern industrial economy had the power to alter the composition of a *Sprachgebiet*. Kiepert thus wanted to have it both ways. He wanted to claim both recently Germanized and the historically German-speaking areas alike as legitimate territorial possessions of the German Empire. Published in the journal of Berlin's Geographical Society, his unofficial boundary map brought the message of the mutability of language borders to an influential audience from the nation's capital.

But over a decade after the German annexation of Alsace-Lorraine in 1871, the French-German language border showed little signs of moving. A second map, surveyed by a Lorrainer man named Constant This in 1888, revealed in stark visual terms the continuing gap between the official and linguistic French-German border (fig. 3.2).[43] On the left-hand side, the map

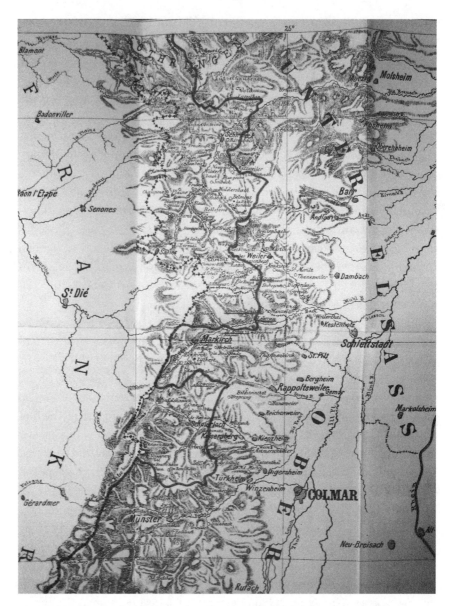

Figure 3.2 Map of the French-German language border, 1888. Note the gap between the language border, indicated with a thick line, and the legal French-German border, indicated to the left with a cross-marked line. Constant This, *Die Deutsch-Französische Sprachgrenze im Elsass* (Strassburg: J. H. Ed. Heitz, 1888). Photo. et coll. BNU Strasbourg.

designated the official border with a cross-marked line, and on the right-hand side, it designated the language border with a thick red line. In the booklet that he wrote to accompany the map, This extolled the scientific merits of his self-surveyed language border. He explained how he had combined knowledge from official language statistics with "direct viewing" (*unmittelbare Anschauung*) and "free observation" (*vorurteilsfreie Beobachtung*) of the landscape.[44] Indeed, in researching the language border, This had spent three months traveling from village to village in the Vosges Mountains in order to identify which villagers spoke a French patois in the privacy of their homes. Like Heinrich Kiepert, he was wary of allowing towns to self-declare their majority language, preferring "objective" social scientists such as himself to make linguistic determinations. In terms of its political impact, This's completed map of the language border—published by a local Strasbourg press—communicated to an educated public of German speakers in Alsace-Lorraine that German nation building in the borderland was still incomplete. The area between the two borderlines on his map symbolized a problematic space where Alsatians had yet to be culturally assimilated into the German Empire. Like the French decades earlier, Germans had begun to grapple with the political implications of their linguistically diverse state.

A LANGUAGE MAP AS REGIONAL PATRIMONY?

While some of Alsace-Lorraine's citizen surveyors focused on the geographic divide between "French speakers" and "German speakers," another group of nineteenth-century mapmakers visualized the borderland's linguistic landscape from an entirely different perspective. During the 1860s, Pastor Louis-Gustave Liebich, raised in Strasbourg and trained at the Protestant seminary there, laid out plans for Alsace's first dialect map.[45] Resisting pressures to categorize language speakers through the lens of nationality, Liebich used his map to valorize the Alsatian dialect as its own independent cultural form. According to Liebich, the Alsatian dialect could be defined as "one of the branches of the great Alemannic trunk that extends through southwest Germany, Switzerland, and our own region [*patrie*]. It is composed of a series of subdialects, each with its special particularities and each attached to a common center with which it shares characteristics."[46] His mapping project constituted an important part of a larger regionalist movement, begun in the 1830s, to preserve Alsace's cultural particularity in the face of an increasingly centralized French state.[47] Like many of his fellow Alsatian bourgeois, Liebich worried that the Alsace of his childhood, where residents of the region's towns and villages spoke variations of *Alsacien*, or *Elsässisch*, was slipping away in the face of industrialization and Parisian meddling. Although he was neither a trained cartographer nor a linguist, Liebich believed that his amateur dialect map could preserve the memory of a time when the melodic sound of *Alsacien* could be heard across the border territory's plains, hills, and mountains. Driven by nostalgia, he hoped that the map could serve, in his words, as a "lasting memorial" to his dying mother tongue.[48]

Like many of the amateur linguistic cartographers in nineteenth-century Europe, Liebich hoped to win the backing of state institutions. Securing financial and institutional support from the French state, however, proved to be a great challenge. Shortly after he announced his map project to an audience of Alsatian readers in the 1861 edition of a popular regional journal, the *Revue d'Alsace*, Liebich wrote a letter to Napoleon III's minister of public instruction, asking the French state to partner with him in distributing language surveys to Alsace's public schools. The language surveys were to serve as the critical evidence base for his map, allowing him to pinpoint the precise locations of dialect boundaries between remote Alsatian villages. Parisian officials, however, refused Liebich's request, calling his idea for a dialect map "impractical" and "irrelevant."[49] Although well aware that there were a large number of dialect speakers in Alsace, the French state preferred to keep knowledge of the nation's internal language divisions private, a strategy first begun by the Napoleonic mapmaker Coquebert de Montbret decades earlier.[50]

Pastor Liebich received a far different response, however, when he solicited German authorities for help with his map proposal after the empire's annexation of the borderland in 1871. Because the German Empire operated as a federated state system, its administration expected and openly acknowledged differences in the form of German spoken across its land. Embracing a far more tolerant view of local dialects than the French, Berlin's newly appointed *Ober-Präsident* in Alsace-Lorraine, the Freiherr von Moeller, threw his support behind the aging clergyman-turned-mapmaker.[51] The German administration in Alsace-Lorraine agreed to distribute Liebich's eighty-question survey of dialect orthography and pronunciation to six hundred schools in Alsace and Lorraine beginning in 1874.[52] To ensure that the dialect information that went into his map represented the "true" linguistic character of each town, Liebich insisted that the surveys were to be filled out by schoolteachers who were native to the village or town in question. If a schoolteacher was an immigrant from another part of Germany, school inspectors were told to find a competent native speaker to fill out the survey, such as a senior schoolboy.[53] Local cooperation was thus at the core of Liebich's cartographic methodology: "It is up to each individual," he wrote, "whether an honorable or a disgraceful monument [to Alsace] will be built."[54]

In their visual form, Liebich's dialect maps reveal his personal interpretation of Alsatian geography, creating an image of a land broken up into a kaleidoscopic array of microcultures (see plate 13).[55] Using hand-painted yellow, blue, green, and red lines, he created "language pathways" (*Sprachwege*, or *Sprachzweige*) that linked together or separated the region's different dialect zones, which he measured against a "baseline" dialect that he selected from his native Strasbourg.[56] Liebich's maps thus supported the developing nineteenth-century theory that dialects were organized in an interrelated geographic continuum.[57] The rare towns whose dialects bore no resemblance to nearby speaking patterns formed "language islands" (*Sprachinseln*, indicated with a star shape), while towns that formed the meeting place between two subdialects formed a "contact ring" (*Verbindungsring*, indicated with squares). Together, the maps' symbols created an innovative portrait of a vibrant, self-sustaining, and complex regional culture.

In rendering visible the invisible linguistic pathways, islands, and contact zones that traversed Alsace-Lorraine, Liebich created what scholars have called a "counter-map." A term first coined by a scholar researching the struggle over land ownership in a former European colony, "counter-mapping" refers to the production of a subversive territorial perspective that challenges the dominant cartographic paradigm of its time.[58] In the case of Liebich's maps, his images presented an alternative view of Alsatian space that emphasized regional culture over national boundaries. Indeed, the audience that Pastor Liebich envisioned for his dialect map was an Alsatian public that was increasingly conscious and proud of its distinct cultural identity. A self-described work of regional scholarship (*Elsässische Wissenschaft*), Liebich's dialect maps reflected a growing sense of confidence among Alsatian regionalists to break open a "third way" between French and German national identities.[59] Even though Liebich's failing health prevented him from publishing his dialect maps before he died, two linguistic researchers, Hans Lienhart and Ernst Martin, quickly acquired the maps and used them as a basis for the cartographic images that accompanied their *Alsatian Dialect Dictionary*, an iconic work of Alsatian regional scholarship published in 1899.[60] Thus, several decades after he first made dialect mapping a public cause, Liebich's images finally found themselves in circulation among Alsatian regionalists, providing them with a comprehensive overview of their unique cultural geography. When Alsace-Lorraine returned once again to French hands, Parisian officials would be faced with the question of how to handle this flourishing Alsatian dialect research, which was helping to promote a strong sense of regional identity among the border population.

RENEWED LANGUAGE TENSIONS ON THE BORDER, 1914–1940

During World War I, when the western front solidified just a few kilometers into Alsace-Lorraine, the language tensions that had simmered in the border area for decades reached a point of crisis. On the defensive militarily, the German government abruptly put an end to the bilingual language instruction that they had allowed in certain schools in Lorraine. In 1915, on the anniversary of Germany's victory over the French at Sedan in 1870, Emperor Wilhelm II ordered all of the region's remaining 247 French village names to be Germanized.[61] In the words of Peter Paulin, a head school-teacher in Strasbourg and an expert on language, war had given "a new face to the western border area."[62] It would be wrong, he argued, to give the German troops who traveled from faraway parts of Germany to defend Alsace-Lorraine a "false impression of the French-language character of the area."[63] Not satisfied that all toponyms on the map of Alsace-Lorraine were German, in 1916, the military authorities also Germanized the names of 1,664 mountains, forests, streams, and other geographic features that still bore French names, in consultation with archive and library directors in Colmar and Strasbourg.[64] Meanwhile, as the Germans fiercely defended their linguistic claims to Alsace-Lorraine, the French mounted their own challenge.

When their army slowly pushed its way deeper into Alsation territory, French officials showed how much their tolerant view of language diversity had changed since Napoleonic times. From 1870 to 1914, the leaders of France's Third Republic had worked tirelessly to teach all of their citizens proper French and to reach a level of linguistic unity that the revolutionaries of 1789 had once sought, and failed, to achieve. During World War I, the French government brought its goal of linguistic uniformity to occupied parts of Alsace-Lorraine. For every new part of the borderland that fell under the French Army's control, commanders opened up French language schools for the population.[65] The military front thus opened the path for France's cultural reconquest of Alsace-Lorraine.

The wartime French occupation administration in Alsace-Lorraine reinforced the notion that the French-German language border was a kind of "national" border through its harsh rhetoric against the German language. German, as Bruno Cabanes, Stéphane Audoin-Rouzeau, and Annette Becker have demonstrated, became increasingly characterized in French wartime propaganda as the crude and uncivilized "language of the barbarian."[66] This characterization, of course, caused immediate tensions with the German-dialect-speaking Alsatian population. In his 1917 text, *Alsatians Correct Our Accent!*, Albert de Dietrich, a Francophile Alsatian of noble roots, condemned his people's continuing use of their Germanic dialect in a wartime atmosphere. "Their bad accent renders them suspect and makes them confused with the enemies," he wrote of his fellow Alsatians, continuing, "How many times have we heard: That's a *boche*, or at least someone from the annexed provinces, which is the same thing."[67] Using the biological metaphor of illness, de Dietrich likened an Alsatian who spoke the local dialect to a sick patient who ignored his doctor's recommendations for treatment.[68]

The language struggles in World War I Alsace-Lorraine thus demonstrated that Europeans no longer considered it enough to claim a border territory based on military might; public opinion demanded a cultural rationale. During the peace negotiations for the Treaty of Versailles, the victors from the Great War redrew the borders of Europe in a manner closely informed by linguistic research, including studies by prominent French geographers.[69] Representatives from the victorious powers, together with nation-building elites from Central and Eastern Europe, used linguistic borders as their guidelines for drawing national borders across the territories of the collapsed Austro-Hungarian, Ottoman, and Russian Empires. The case of Alsace-Lorraine deviated from this pattern; as the victors, the French were in a position to dismiss the significance of the French-German language border in the Vosges Mountains and relocate their eastern border along the midpoint of the Rhine River, where it had stood from 1648 to 1871. They did not, however, ignore the reality that Alsace-Lorraine was majority German-speaking. French authorities in the postwar period were determined to strengthen their cultural claims to Alsace-Lorraine by turning the borderland into a French-speaking zone. They immediately sent hundreds of French language teachers to Alsace-Lorraine from across France, including regions as far away as Brittany.

To monitor the progress of the French language in the borderland, the French census bureau made a dramatic break with tradition and decided to gather language statistics for the first time since the Napoleonic period, exclusively for Alsace-Lorraine, beginning with the French National Census of 1926. The French census takers set up their language categories in such a way, however, that it enabled Alsatians and Lorrainers to self-identify as dialect speakers rather than as German speakers.[70] The census of 1931, for example, gave individual responders nine categories from which to choose: French only (11%); French and dialect (4%); French and German (8%); French, dialect, and German (30%); dialect only (6%); dialect and German (27%); German only (6%); other languages (1%); and not declared (7%).[71] The purpose of these numerous categories was to use the presence of dialects in Alsace-Lorraine to diminish the appearance of a German-speaking population. Motivated by border tensions with Germany, the French census bureau framed the Alsatian dialect as a distinct language that was one of France's many regional patois, denying its status as a linguistic branch that grew from a German tree.

In September 1940, the cultural politics of the Alsatian dialect came full circle. Shortly after Germany's invasion of France, the French state executed for sedition a former scholar of the Alsatian dialect and the leader of the pro-German Alsatian autonomist movement, Karl Roos.[72] During the turn of the century, Roos had studied at the University of Strasbourg under Professor of Linguistics Ernst Martin, one of the coeditors of the *Alsatian Dialect Dictionary*, to which the regionalist mapmaker Louis-Gustave Liebich's research had contributed. Recognized as a promising young scholar of Alsatian linguistics, Roos published his doctoral dissertation, "Foreign Words in the Alsatian Dialects," in 1903.[73] While Roos later chose to focus his career on politics, his study of the Alsatian dialect no doubt had a formative influence on his strong sense of regional identity. Under fire from French authorities during the interwar years, the Alsatian dialect came to represent for the autonomists a symbol of their cultural freedom and pushed them toward an alliance with the Nazi Party. Shortly after the second German annexation of Alsace-Lorraine, the German occupying authorities rechristened Strasbourg's central square, the Place Kléber, "Karl Roos Platz."

The language borders that European mapmakers invented during the nineteenth century carried no direct jurisdictional authority, nor were they visible to the naked eye with stone border markers. When they entered the realm of cartographic space, however, language borders became powerful and real. They presented Europeans with a system of borders that was an alternative to the legally negotiated and scientifically surveyed land divisions of the past. Above all, language borders provided a framework for thinking about European territory in terms of a human geography of nationally or regionally defined cultural groups. Rarely did these cultural groups align with existing state boundaries. Napoleonic-era cartographers discovered "German" communities in France's eastern departments, and German nationalists faced a problematic area of French speakers in Lorraine. Alsatian regionalists invented a utopian dialect space that defied state borders altogether. While many twentieth-century European governments adopted policies of nationalizing their state borders—"purifying" their borders through massive ethnic population transfers and swaps in the 1920s—the nineteenth century had been a crucial time for discovering and mapping the disconnects between state and national spaces. The twentieth-century "nation-state" unravels when traced back to the nineteenth century; "nations" and "states" were not one and the same, but separate entities, mapped with different techniques.

To map the boundaries of language during the nineteenth century, Europeans turned to creative methods. Unlike the demarcation of state boundaries, mapping language borders never required the intensive labor and significant expense of government-sponsored border commissions. To the contrary, individual mapmakers from all corners of society, both inside and outside the state, developed their own highly individualized methodologies and aesthetics. Relying on cultural knowledge rather than on topographical measurements, linguistic mapmakers understood the power of manipulating the categories of information that they used, particularly labels such as "dialects" versus "languages." While some language mapmakers experienced the cultural terrain of Alsace and Lorraine firsthand, walking across the border zone and observing the local population as field researchers, others simply drew their maps from their offices in Paris or Berlin. No matter the techniques, however, the art of surveying a language border and fixing it in space required a combination of imagination and political will. This is why historians must interpret maps of language borders as arguments about cultural identity rather than as mirrors of reality.

The historical value of modern Europe's language map archive, however, goes well beyond the social and cultural information locked up inside its paper images. In analyzing how people circulated, used, and displayed language maps, we can learn how the images shaped public views of borders. Language maps became a powerful rhetorical device for both nationalist and regionalist claims to territorial space in Europe. Thanks to the invention of cheap printing technologies, members of the European public not only gained unprecedented access to looking at maps; they also acquired the newfound ability to invent and produce their own types of politically motivated map images. Part II of this study continues our investigation into the links between popular mapmaking and identity politics in Alsace and Lorraine. Our focus will shift, however, from the ordered and "rational" visual language of social scientific maps to a new set of mapping techniques designed to stimulate citizens' emotional and sensory connections to their land.

Borderland Maps for Everyday Life

CHAPTER 4

Finding the Center

Maps from cultures across the world teach us that the feeling of home is closely tied to the spatial idea of the center.[1] For European governments and armies, as well as the leaders of European nationalist movements, capital cities provided the ideal political and cultural beacons for orienting faraway border provinces around a unified national territory. But there was a problem with maps that framed the "homeland" as a space that flowed outward from capital cities. Many provincial Europeans, especially those living in remote border regions, had never seen their state capitals with their own eyes. In disputed Alsace and Lorraine, most people found it far easier to grasp their spatial bearings, and their sense of cultural and national identity, through maps centered around the places that were closest and most familiar to them: their hometowns. Unlike the single state center symbolized by the scientific observatories in Paris or Berlin, hometowns provided citizens with spatial reference points that they knew intimately, through their daily experiences. For the majority of Alsatians and Lorrainers, their hometown was a medieval village with a single church spire and a cluster of half-timbered houses, surrounded by a landscape of vineyards or forests.[2] During the nineteenth and twentieth centuries, these picturesque villages became fertile ground for the development of innovative visual strategies aimed at building the border population's loyalties to the region and the nation through its interaction with its local environment. It was a radically different philosophical approach to national integration in which the local hometown, rather than Paris or Berlin, became the center around which the rest of the nation radiated.

Comparing local and regional maps of Alsace and Lorraine from periods of French and German rule, we see that both countries used cartographic strategies to establish relationships between their territorial parts and the whole.[3] For Germans, the concept of home, or *Heimat*, could simultaneously refer to one's native village, province, and nation; the word does not exist in the plural.[4] On map images, such a flexible definition could expand or contract to accommodate one's local, regional, or national "home" (or a combination of the three). In France, nationalists similarly used maps to build a bridge between a citizen's tangible "little homeland," or *petite patrie,* and the more conceptual "big homeland," or *grande patrie.*[5] While recent studies on French and German nationalism have emphasized the interactive nature of local, regional, and national identities, the visual relationships among hometowns, regions, and the nation have gone largely unexplored.[6] Examining how subnational spaces appeared on French and German maps can therefore teach us a great deal about the similarities and differences between multiple levels of territorial identity in the two countries.[7]

To train ordinary Alsatians and Lorrainers to think of their familiar and beloved hometowns as inherently "French" or "German" spaces, nineteenth-century cartographers turned to a number of creative visual techniques. Perhaps most importantly, they drew their local maps on vastly larger scales (1:2,500) or (1:5,000) than most scientific or social scientific maps of France or Germany. Choosing a large scale meant that a mapmaker could highlight the local cultural landmarks that villagers were used to seeing with their own eyes: their schoolhouse, their town hall, their farm fields, and their village bell tower. In collapsing the gap between "experienced" and "abstracted" space, local maps became incredibly persuasive objects. While appearing to mirror the places that villagers looked at and traveled through in their daily lives, local maps in fact manipulated their appearance by erasing or emphasizing particular cultural landmarks.

In addition to their large scales, local maps returned to the kind of old-fashioned pictorial imagery that modern survey maps and national atlases, with their scientific and rational modes of representation, had shed. To a certain extent, their visual language harkened back to chorography, a traditional form of early modern European mapmaking that focused on regional topographical description.[8] Like the older chorographic maps, these nineteenth- and twentieth-century regional maps overflowed with soft, delightful images of local flora, fauna, produce, and heroic figures, conjuring up a warm, intimate feeling of home. Designed to tug at the heartstrings and elicit passionate feelings of local pride, the images were emotionally, as well as visually, manipulative. Through the visual language of the local map, ordinary places became utopian places—idealized images of nationally rooted villages that fitted within a constellation of "French" or "German" hometowns.[9]

Local maps could connect successfully with the experiences and worldviews of ordinary Alsatians and Lorrainers in part because people fashioned them from the "bottom up." Civilians were eager to make use of the new visual technologies available to them in their struggle to define their identities within larger regional and national communities.[10] In Alsace-Lorraine, local schoolteachers, rather than central governments, developed many of the lessons for primary and secondary school geography curricula. These purveyors of local knowledge often spoke the Alsatian and Lorrainer dialects and were intimately familiar with their hometown culture, architecture, environment, and history. In addition to teachers, individual artists, such as the famous Alsatian cartoonist Hansi, and private associations printed and circulated so many visual depictions of rural villages that the aesthetic category of the "local image" (*image locale*, or *Ortsbild*) came to dominate a great many print magazines and public exhibitions in nineteenth- and twentieth-century Alsace-Lorraine.[11] The scale of production and distribution of these local maps and images was especially remarkable given that "the homeland" was a term that never carried any official jurisdictional status. It was an emotionally charged space with ill-defined boundaries that existed only in the minds of the public.[12] But, as we shall see, the ambiguous spatial definition of the homeland was in many ways a great asset to nationalists seeking to make territorial claims to Alsace-Lorraine.

In spite of their usefulness to nationalists, however, regionalists often appropriated maps of local space for a different political purpose. Local mapmaking could easily transform into an act of cultural resistance through the construction of what some scholars have called "fugitive landscapes."[13] Rather than connect map images of their region to a larger image of France or Germany, Alsatian regionalists sought to valorize their region as a self-sustaining, stand-alone space.[14] To do so, however, they had to resort to their own homogenizing visual strategies, making local and regional landscapes appear less complex by glossing over particularities within the region itself. Focusing on Alsatian examples, leaving a discussion of regionalism in German-annexed Lorraine to other authors,[15] we will examine how the fracturing of national territory into small spaces, both local and regional, could come back to haunt nationalists in modern Europe.

CADASTRAL MAPS: OFFICIAL PORTRAITS OF LOCAL SPACE

Because the homeland was an emotional concept, it was not a place that states could demarcate as an official, clearly bounded territorial unit. Both French and German states did, however, create a body of government-sanctioned visual representations of Alsatian hometowns in the form of cadastral maps, and it is with these maps that we begin our investigation into the genre of local cartography. While the primary function of cadastral maps was to serve as fiscal documents for determining the value of properties within Alsatian townships, we can also analyze them as iconographic images with their own set of visual politics.[16] The decorative imagery on Alsatian cadastral maps visually communicated the relationships among locality, region, and nation that the French and German governments promoted during their rule. Cadastral surveys in Alsace date to the Napoleonic period, and the French government completed its first complete set of Alsatian village plans (*tableaux d'assemblage*) in 1828.[17] Alsatian surveyors hand drew the maps on scales of 1:20,000, 1:10,000, or 1:5,000, much larger than the scales of the military's topographical maps of the region, so that the property boundaries of each village landholder could be clearly visible. Following their annexation of the borderland in 1871, it took the Germans over a decade to revise the French cadastre.[18] When the German state finally introduced its cadastral law to Alsace-Lorraine in 1884, it undertook a long-term project to revise the antiquated French cadastral maps. In addition to enforcing a new tax system, the new German cadastral maps of Alsace and Lorraine established a uniform and elaborate visual iconography—a marked change from the maps' sparse and simple decorative elements from the French period—that illustrated the developing relationship among local, regional, and national identities within the recently founded German Empire.

During the last two decades of German rule in Alsace-Lorraine, each of the new village plans (*Übersichtskarte*) printed at the cadastre office in Strasbourg indicated the name of the pictured village within a cartouche (text box) that contained a decorative image of the Alsatian countryside.[19] The illustrated scene, rich in symbolic details, requires the historian's close inspection (see fig. 4.1a). Pictured on the left side of the cartouche is a surveying instrument perched on a tripod, a reference to the technical expertise that went into creating the map. Directly below the surveying instrument is a shield that is intentionally left blank so that each municipality could add its particular coat of arms. The vegetation displayed in the image includes local Alsatian plants and foliage, notably grapevines and huckleberry bushes. In the right-hand corner of the image sits a brawny figure that appears to be from a bygone age: sitting barefoot, he is wearing the traditional clothing of an agricultural worker, seated on part of a plow, holding his scythe. He looks out into the distance at a pastoral scene that contains a small village with a church and rolling mountains.

The man's commanding gaze over the picturesque scenery suggests his close connection to his native village and its surrounding land. He is the epitome of a European farmer who knows his village's spatial boundaries by heart and through his feet.[20] However, there is a trick to the image. While the map cartouche appears to depict Biblisheim's unique surroundings, it is in fact the government's rubber-stamped, stereotyped rendition of a rural Alsatian scene that is identical for all of the villages in the region (compare

Figure 4.1a German cadastral map cartouche from the village of Biblisheim, 1909. Archives départementales du Bas-Rhin, series Plan/2P.

Figure 4.1b Wide view of the Biblisheim map. Archives départementales du Bas-Rhin, series Plan/2P.

Figure 4.2 German cadastral map cartouche from the village of Fröschweiler, 1911. Note the iconography identical to that in the cartouche on the Biblisheim map. Archives départementales du Bas-Rhin, series Plan/2P.

fig. 4.1a to fig. 4.2, an image taken from the cadastral map of Fröschweiler, a different Alsatian village). The mass-produced cartouche presents a utopian vision of the border province that is frozen in time, without factories or any signs of the industrial economy that was at that time making heavy inroads into Alsace-Lorraine. With the exception of the shield onto which individual villages and townships could add their coats of arms, my extensive archival survey confirms that this iconographic image was identical for all of the cadastral plans from the different administrative circles (*Kreise*) in Alsace-Lorraine.

French cadastral law returned to Alsace-Lorraine in 1923, five years after the borderland's return to France, but the imagery on cadastral maps maintained some striking continuities with the German period. In fact, at first glance, the cartouches for the French cadastral maps appear identical to the turn-of-the-century German ones (see fig. 4.3 for an image of the cartouche from Hermerswiller's cadastral map). Only upon closer observation do a series of subtle but significant changes become visible. The man seated in the right-hand corner of the image is wearing the traditional clothing of an agricultural laborer, but his stance appears more relaxed and his body ap-

Figure 4.3 French cadastral map from the village of Hermerswiller, 1923. Note the subtle changes from the iconography of the cartouches on the German cadastral maps. Archives départementales du Bas-Rhin, series Plan/2P.

pears less masculine than the man in the German cartouche, particularly his arm and thigh muscles. His covered feet suggest a civility that is lacking in the barefoot German figure, and his scythe is now turned downward, suggesting a less aggressive posture. In spite of the differences between the two human figures, however, the fact that both French and German mapmakers chose to emphasize male personages, rather than well-established female representations of the nation such as Marianne or Germania, tells us that there was something about agricultural production that both cultures symbolically associated with maleness.

There are also key differences between the Alsatian landscapes pictured on the French and German cartouches. The French man sets his gaze upon a rural scene that evokes the same calm, pastoral air as the German cadastral map, but the terrain has been transformed into a smoother, flatter landscape reminiscent of central France. There is also a significant change in vegetation. Behind the seated figure, the grapevines from the German image are replaced by a fir tree, a prominent cultural symbol for the "lost provinces" of Alsace and Lorraine popularized by the influential turn-of-the-century French classroom reader *Le tour de la France par deux enfants*, which I will discuss in detail later in this chapter.[21] What about the blank shield from the German cartouche? Revealing a particularly French understanding of the relationship between locality and nation, the white space that the Germans had reserved for village seals is inscribed with the official coat of arms for the French Republic with its revolutionary fasces, a symbol of republican unity, and the Phrygian cap of the revolutionary citizen. Read as a political document, the iconography of the interwar French cadastre signified a break from the federalist and more localized patterns of government under the German Empire and a turn toward a government system supported by a more centralized conception of political identity.

The French and German cadastral maps were exposed to a wide viewership. Under the German Empire, village mayors could purchase them for 5 or 8 marks, depending on their size.[22] The French, in turn, offered the cadastral plans for purchase at a price of 4.5 or 7.5 francs.[23] The cadastral maps were not only used for property sales or land disputes within particular villages; correspondence between the educational administration and the cadastral office in Strasbourg reveals that village schools commonly used them as a visual references for classroom geography instruction.[24] In view of their diverse uses within municipal communities, it makes sense for historians to examine the political concepts expressed in the cadastral maps as part of the French and Germans' larger nation-building goals in the border region.

For example, it is significant that both German and French cadastral maps from the early twentieth century situated all of Alsace-Lorraine's village communities within an identical "regional" mold. On both the French and German sets of cadastral maps, the regionalization of the Alsatian village was achieved first through visual abstraction (leaving divisive or controversial aspects, such as Protestant, Catholic, or Jewish religious identity, out of the image of the farmer and his village) and, second, through visual repetition (inventing a "type" of Alsatian village by creating an identical portrait for each of Alsace's diverse rural communities). The continuity between the "regionalization" strategies of the German and French cadastral maps demonstrated the strength of regional identity not only in Germany, traditionally recognized for its federal political organization and cultural diversity, but perhaps more surprisingly in Third Republic France, a far more centralized state. French and German governments may have tolerated regional and local political structures to varying degrees, but both had to accommodate the persistence of subnational territorial attachments.

As government-approved icons of Alsatian villages, the cadastral maps offered soft, familiar, and welcoming views of the border region that differed fundamentally from the unemotional territorial perspectives that appeared on scientific and social scientific maps. Their iconographic images embody Alon Confino's description of a typical *Heimat* (homeland) image: a small human settlement with a church tower set in a landscape populated with simple people, the viewer often perched behind a faraway hill, creating a visual image that was "warm and desirable, not sublime but friendly and intimate."[25] To understand the origins of the Alsatian villages' particular iconographic representation, we must situate the cadastral maps within a broader corpus of local and regional images that emerged from both state institutions and the Alsatian public sphere during the late nineteenth and early twentieth centuries. One of the richest collections of village imagery comes from the maps, photographs, and lithographs that village schools displayed in geography classes. Together with the Alsatian village iconography created for commercial publications, these classroom images helped ordinary people to mediate between their tangible local reality and the abstract concepts of region and nation.

OBSERVING THE HOMETOWN AND UNDERSTANDING THE WIDER WORLD

Schools provided a place for German children to learn about their young empire's developing political order. After unification in 1871, the imperial government struggled to develop a nation-building strategy "from above": officials in Berlin could not even manage to establish national holidays, a national anthem, or a national flag. Over time, imperial administrators came to view local and regional identities as the most viable conduit to building a German national consciousness out of a union of strong federal states. To facilitate nation-building "from below," teachers and small printing shops in Alsace-Lorraine developed classroom maps and geography texts that focused on local cultures and spaces. A relatively underdeveloped subject until the late nineteenth century, the discipline of geography came to play an important role in reshaping the rural public's perception of local spaces in order to popularize the idea of Germany as a nation of "a thousand *Heimats*"[26]: a harmonious union of diverse homelands that had come together as a nation to promote their shared interests. While certain domineering images of the German Empire, such as imperial maps or portraits of the kaiser, hung on classroom walls, many of the images present in turn-of-the-century Alsatian schools pictured local sites that stood a stone's throw from the schoolhouse: the church, mill, town hall, post office, local streets, and railroads. This approach to teaching German patriotism through geography, however, was more difficult to establish in Alsace-Lorraine than in other German regions because of the newness of the *Heimat* concept in the previously French territory and the region's abnormal status within the federated German Empire as a special imperial territory with limited political rights.

In order to orient students effectively in their local geography, the first lesson in Alsace-Lorraine's new German geography curriculum was an outdoor walking excursion through the students' village and its surroundings. Alsatian schoolteachers referred to the field trips as exercises in "active observation" (*lebendige Anschauung*), designed to instill in the child's mind a "homeland image filled with life" (*lebensvolles Bild der Heimat*).[27] This pedagogical technique was grounded in the philosophy, dating back to the Enlightenment and Jean-Jacques Rousseau's *Émile*, that knowledge should be acquired through sensory experience.[28] Using their senses of sight and touch, students learned about the fundamental principles of geography—plant and animal life, soil composition, topographical forms, and directional orientation—from physical evidence available in their hometown. A geography lesson plan, modeled on a text from neighboring Baden and finalized at an 1872 teachers' conference in Strasbourg, described the atmosphere of a class field trip:

> We turn our attention to the plants, flowers, and herbs; to the trees and vines that grow here; to the birds and other animals that live here. We also investigate whether the ground is sandy, clay-like, marl-like, or stony, and, in the latter case, what kind of stones they are. When we arrive in the next town, we stop for a little while at our neighbors', to get to know them better and to see what they do. We also observe their church and their school, to be able to compare them.[29]

The empirical comparison between the students' hometown and its surrounding villages reinforced the notion of local particularity. Even the map image that accompanied the outdoor exercise demonstrated that the goal of early-year geography education was to provide students with a sense of local importance: "We draw our hometown with a big point in the middle of the board," explained the teachers' manual. "Town A, which against the sunset lies to the right; the *Meierhof* B, to the left; village C, and mill D, under."[30]

During the 1880s and 1890s, when the concept of *Heimat* emerged as a powerful alternative to Berlin-directed, imperial nation building, local village tours became increasingly politicized. What changed with the introduction of *Heimat* ideology into school geography curriculum was the connection between local pride and German national identity. Along with history and music, geography became an academic pillar of *Heimatkunde*, a genre of local cultural studies popular among the nineteenth-century German bourgeoisie.[31] In turn-of-the-century Alsace-Lorraine, the teachers who wrote elementary school geography textbooks, including German immigrants and native Alsatians alike, were self-proclaimed *Heimatler,* or "wise men" of local history and culture. Even nationally circulated geography books, such as Johan Ludwig Algermissen's mass-produced altases, were printed locally, in small publishing houses, with a section set aside at the beginning of the text for a contribution from a local expert.[32] These local experts, patriots invested in the German national project, transformed Alsatian villages into museums of cultural landmarks. They taught children to be proud of their village's particular terrain, soil, waterways, buildings, climate, and population but also to expand their local pride to a feeling of affection for the whole German Empire.

During this flowering period for local studies, geography books used innovative visual strategies to train Alsatian children to see their village surroundings through the gaze of the *Heimat* concept. "The Hometown" (*Heimatort*), an 1894 lesson designed by Georg Weick, a head teacher in Saarbourg, is an example of one such text that fused local pride, visual imagery, and geography.[33] Weick instructed teachers to begin their lesson by introducing students to the empirical methods of geography: students and teachers worked together to draw a to-scale map of their classroom on the blackboard with the aid of a measuring stick. The teacher placed a compass on floor of the classroom and helped the students to add the markings for north, south, east, and west to their map of the schoolhouse. Then the teacher drew a new map on the blackboard, where the schoolhouse was shrunk and bordering streets were drawn to scale according to measurements taken by the students themselves, who were free to leave the classroom to measure the streets' distances by counting their paces. The students took similar outdoor measurements even further away from the schoolhouse in order to incorporate their hometown's main streets, squares, the creek, important buildings, monuments, the railroad, and the canal into the map image.

Actively engaging with the map and personalizing its symbols, students copied the blackboard's "master" village plan onto their individual slates, adding where they lived, and marking the route from their home to the school, the church, the post office, and the railroad station. The teacher then asked them to gauge the size of their town by drawing the first and last house in the hometown, adding the homes of "important people" as directional reference points. They also had to determine the shape of their hometown from a planimetric view: "How does our hometown appear on the map image? Is it something like a cross, a triangle, or a star?" asked the teacher.[34] Students further investigated the hometown's physical shape on a field trip to the highest point in the surrounding countryside, from which the geography class could acquire a panoramic view. From this position, the teacher pointed out the location of his students' village with respect to other villages, using church towers as reference points to compare the villages' heights and shapes.

In the last part of the lesson, Weick emphasized the relationship between the map on the blackboard and the students' emotional attachment to their local homeland. Upon returning from their field trip to the panoramic viewpoint, the teacher led his students in singing a celebratory song: "It is beautiful in the homeland / On the mountains' rising heights / On the craggy cliff paths / On the meadows' green crops / Where the herds go to graze / It is beautiful in the homeland, it is beautiful in the homeland!"[35] Folk songs such as "It Is Beautiful in the Homeland" (In der Heimat ist es schön) reflected conceptions of place and identity that were well rooted in Alsatian village popular culture.[36] The songs were traditional expressions of local pride, many of them more than a hundred years old, which infused the natural environment with emotion and sentimentality.[37] Their vivid imagery expressed a love for the particular local spaces in Alsace; one specialist in Alsatian folk songs even referred to the songwriters as *Bilderjäger*, or image hunters.[38] Popular songs thus formed an important conceptual bridge between a well-rooted folk culture and the emerging modern concept of the *Heimat*, which repackaged local images and connected them to a bigger national space.

Rather than provide a synoptic view of German national territory, localized geography instruction supported a method of visual training that encouraged children to conceptualize large territorial spaces from the locality outward. "In geography and history classes, localities and the region of Alsace-Lorraine predominated, but they were at one end of a continuum. At the other end was the German nation," explains Stephen Harp, author of a study on Alsatian schools.[39] First, students learned to orient themselves in their immediate settings, familiarizing themselves with the four cardinal directions and the processes of local mapmaking. Then they would use real-life orientation exercises to conceptualize an abstract imperial territorial space that they could never see or experience in its entirety. But between experiencing their village topography on field trips and visualizing the German Empire, there was a key intermediary space. Germany was not just a collection of thousands of hometowns; it was a family of regions (*Länder*). Wiped off the map and converted into administrative departments in 1790, Alsace and Lorraine had not existed as official territorial units since Old Regime France.[40] During the 1880s and 1890s, a modern form of regional consciousness developed in Alsace-Lorraine that built upon the seeds of regional identity that had developed earlier in the nineteenth century under French rule. An important step toward strengthening Alsatian and Lorrainer regional consciousness was to train students to locate their hometown on regional maps.

FROM EXPERIENCED TO ABSTRACTED SPACE: SITUATING THE HOMETOWN WITHIN THE MAP IMAGE OF ALSACE-LORRAINE

Maps played an important role in introducing the Alsatian population to the concept of regional identity, symbolically bringing together the men and women from the borderland's diverse villages, townships, and cities into a single community. Geography teachers introduced schoolchildren to the concept of the region by placing their familiar, nearby *Heimat* within the paper image of a regional administrative map. "The maps of the canton, the district, as well as the neighboring Fatherland must be deeply imprinted with the map image of *Heimat*," explained one geography teacher.[41] Maps displaying units of local government were useful to schoolteachers because their scale was large enough to contain even the smallest Alsatian villages. They therefore helped students to make the connection between their "experienced space" (their hometown and its surroundings) and Alsace-Lorraine's "abstract regional space" (which was too large to know through experience and could only be understood with the aid of a map). As the schoolteacher Georg Weick explained, "After an understanding of maps has been achieved from reality, an understanding of reality can be achieved through maps."[42]

Beginning in the 1870s, an Alsatian-based publisher, the Boltze Book Company, printed a series of colored wall maps for each of Alsace-Lorraine's administrative circles (*Kreise*). A German-born entrepreneur who moved to the Alsatian village of Gebweiler a few months before the Franco-Prussian War, Julius Boltze dedicated his printing company to strengthening civic life and improving practical training in Alsace-Lorraine.[43] Boltze printed his classroom wall maps, a simplified version of the government's district maps, on a large scale with bold colors and decorative symbols for local landmarks (see fig. 4.4). They indicated towns, mills, churches, rivers, streams, railroads, streets, cities, and administrative divisions. Boltze's classroom maps offered students a powerful opportunity to orient themselves in geographic space: students could examine the map and search for their particular hometown within a territorial unit of imperial administration. Imperial authorities presented Boltze with an educational award for the success of his district maps, praising them for their contribution to *Heimat* studies.[44]

In their study on Swiss cartography, historians David Gugerli and Dan-

Figure 4.4 During the turn of the century, classroom wall maps became an important tool for grounding abstract ideas of nation, region, and locality in territorial space. François Xavier Saile, *Wandkarte des Stadt- und Land-Kreises Strassburg* (Gebweiler: J. Boltze und R. Schultz, 1876). Photo. et coll. BNU Strasbourg, M.Carte.175.

iel Speich discuss the emotion and excitement that Swiss townspeople felt upon locating their hometown on the national map of Switzerland displayed for the first time at a public exhibition in the mid-nineteenth century.[45] Gugerli and Speich point to the connection between the townspeople's experience of "self-orientation" on the national map and a growing consciousness of national identity. Boltze's district maps offered this opportunity on a regional scale: the student could place him- or herself within a collectivity greater than his or her village community through the subtle act of locating his or her hometown within the district map image. "Every teacher and student will be pleased when they find their hometown," explained one Alsatian textbook, in reference to the wall-sized district maps, "they know from experience how far it is to the closest town; they find this distance on the map and gain a better sense of scale for the area that is represented."[46] When students found their hometown on the district map, they could then relate an abstract visual representation of space (the map image) to a known environment (the hometown and its surroundings).

Besides situating the hometown within the territorial units of the German government, Boltze's classroom wall maps placed the hometown within a growing regional railway network. The railroads displayed on Boltze's maps suggested that Alsace's hamlets and towns were linked together in a network of regional association, offering citizens the opportunity to visualize their region as a coherent whole. While historians have emphasized the role of railroads in bringing together national publics, less attention has been paid to railroads as the means for constructing regional spaces.[47] During the late nineteenth century, for example, Alsatians used trains for local purposes more frequently than they did for long-distance journeys; residents of Strasbourg traveled to the Vosges for leisure, and rural folk traveled to Alsatian cities for their economic and administrative affairs. Even today, railways in European countries are still divided into separate regional and national networks, the former usually slower and cheaper than the latter.

Situating the hometown and its local transportation network on a district wall map, however, was just the beginning of a student's introduction to the core subnational territorial units within the German Empire. To familiarize children with the visual shape of their region, schoolteachers taught their students how to create their own, hand-drawn map images of Alsace-Lorraine, similar to the maps they drew of their hometown. A textbook published in 1875 taught students how to draw the borders of Alsace-Lorraine and the German Empire in order to train them to "see geographically" and to help them memorize the form (*Gestalt*) of their fatherland (see fig. 4.5).[48] Working off sheets of tracing paper that pictured the borders of each territorial unit, students repetitively drew the contours of Alsace-Lorraine, the Vosges, and the German Empire until they were familiar with their particular shapes. The goal of such exercises was to create "a clear image that [was] easily impressed upon the memory."[49]

The purpose behind the simple visual images of region and state used in Alsatian schools was best explained by an unlikely figure: Martin Kunz of Mulhouse, a well-known European expert on pedagogy for blind children. Kunz believed that blind children should be provided access to their society's key visual concepts. In his 1891 lecture delivered before the German Association of Teachers for the Blind at Kiel, Kunz discussed ways to help blind children to fit into the dominant seeing culture. "Images," he remarked, "are small joys and are also a treasure trove of countless ideas

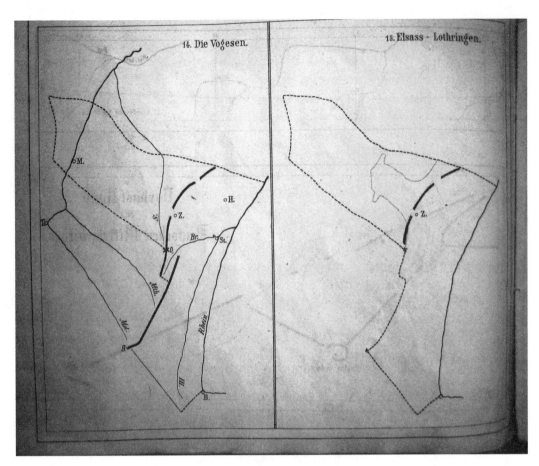

Figure 4.5 Students in Alsatian and Lorrainer schools used sheets of tracing paper to memorize their region's shape. Image from G. Moser and G. Kaufmann, *Geographische Faustzeichnung als Grundlage für einen methodischen Unterricht in der Geographie* (Strassburg: R. Schultz u. Comp., 1875). Photo. et coll. BNU Strasbourg.

that serve as the building blocks for the development of [children's] spiritual lives."[50] In his *Atlas for the Blind* (*Blinden Atlas*), printed through the support of the Institution for the Blind in Illzach (southern Alsace), Kunz stamped curved strips of braille onto white paper to help blind students create a mental image of several key territorial concepts that represented the continuum from region, to nation, to the wider world: the atlas's images included Alsace-Lorraine, the German Empire, and other countries in Europe (fig. 4.6).[51] Rising and falling contours within the dotted strips of braille were intended to familiarize blind students with topographical variations in terrain through their sense of touch. Some of his maps included grooves for railway networks. Thanks to Kunz's enterprise, his Alsace-based institute successfully fulfilled its mission, selling thirty thousand copies of his "relief maps" to the German Association for the Blind, and an additional thirty-three hundred maps to other institutes for the blind across Europe.[52]

Why would teachers of the blind and sighted alike go to such lengths to

Figure 4.6 This braille map of Alsace-Lorraine was part of an atlas designed to provide blind children with a foundation in society's key visual concepts. Martin Kunz, *Blinden Atlas*, 1891. Fonds patrimonial des médiathèques de Strasbourg.

provide their students a mental image of Alsace-Lorraine's physical shape? When students learned to visualize their region in space, the teachers believed, they would begin to think of the German Empire as a family of regions like theirs. But memorizing Alsace-Lorraine's physical contours would not be enough to instill a sense of regional pride in Alsatian schoolchildren. In order to strengthen their students' emotional attachment to their region, teachers needed to figure out a way to place maps in dialogue with other kinds of visual images. A renaissance in regional iconography, made possible through technological advances in lithography and photography, helped schoolteachers to enrich their lessons on Alsatian geography.

CANONIZED IMAGES AND THE VISUAL NARRATIVE OF REGIONAL IDENTITY

In medieval Europe, the architects of Gothic churches assembled stained-glass windows in a particular sequence to provide an illiterate population with a simple and direct narrative of Bible stories. Similarly, Alsatian region builders constructed an identity out of a small number of expressive visual reference points, condensing a complex regional experience into a select number of canonized images. Late nineteenth-century Alsatian geography texts were filled with mass-produced lithographs and photographs of Alsace that promoted a narrow view of the region as an idyllic land of half-timbered houses, vineyards, church steeples, and storks' nests. These images transcended the particularity of individual Alsatian villages, creating a universal *Ortsbild*, or local image, of Alsace-Lorraine. The term *Ortsbild* refers to a landscape that has been turned into a type or a caricature; it is an image that always contains the same elements and has a formulaic quality.[53] One can think of the Alsatian local image as a visual archetype that smoothed over differences within the region, particularly the historical tension among Alsace's Protestant, Jewish, and Catholic communities, to produce an ecumenical representation void of religious symbolism.[54] Further, the *Ortsbild* concept offered a means for the people of Alsace-Lorraine to develop the same strong regional identities that already existed among the people of the other German states, who had enjoyed autonomy during the centuries that Alsace and Lorraine had been under the grip of Paris. The *Heimatlers'* utopian images of Alsace-Lorraine were thus analogous to those created by nation builders; they presented uncomplicated categories of identity based on generalized cultural tropes. Turn-of-the-century Alsatian regional iconography—"the Alsace of little intact villages"[55]—created a standardized mold into which any Alsatian village could fit.

The mass production and widespread dissemination of regional images was linked to nineteenth-century innovations in visual technology. Beginning in the 1880s, there was a sudden profusion of photographic images in upper-year geography textbooks in Alsace-Lorraine. These images were critical to creating a simple yet vivid visual narrative of Alsatian identity; they acted in tandem with regional maps to promote a cohesive sense of regional togetherness. One geography text, for example, printed a set of photographs reminiscent of the picturesque postcard images sold in the region's tourist industry. The text included images of a diverse group of landmarks (*Wahrzeichen*) in Alsace-Lorraine, including the Cathedral of Notre Dame and the Kammerzell House in Strasbourg, the Town Hall of Mulhouse, the Castles of St. Ulrich and Hoh-Königsburg, a lake from the Vosges Mountains, and an ethnographic-style photograph of a *Schlitteur* (the term for a logger from the Vosges).[56] Accompanying these photographs was an image of the train routes that bound the towns of Alsace-Lorraine together into a single spatial unit. The railroad image suggested a "regional itinerary" to go along with the collection of urban and rural visual signposts provided by the photographs. Schoolchildren in Alsace were thus presented with the same polished, idealized view of their region as tourists.

Some textbooks for upper-year geography students embraced a form of instruction called *Landeskunde* (regional studies), an expanded form of *Heimatkunde* (homeland studies) that strived to connect the diverse microenvironments in Alsace into a single regional community. The images appearing in these texts were generally more ethnographic than picturesque. One text written by a schoolteacher in Strasbourg offers a good example.[57] Turning an anthropological eye on the borderland, the text's lithographic images distilled the Alsatian population and their culture into a series of "types." One page portrayed the inhabitants from Alsace-Lorraine at work on the land, chopping down trees and burning wood chips. Such an image symbolically tied the Alsatians to their land; it was a nostalgic vision of people who had not yet abandoned rural Germany for the factory. The other side of the page pictured three women and a man from Upper Alsace, Lower Alsace, and Lorraine in traditional folk dress (fig. 4.7). The image politicized the clothing worn by the people of the region, turning the particular Alsatian and Lorrainer clothing styles into symbols of regional identity. A second page showed two of the most common housing styles in Alsace-Lorraine: Frankish and the Allemanish (fig. 4.8). Regional housing

Figure 4.7 Ethnographic images of work life and clothing in Alsace-Lorraine. Reproduced from E. Rudolph, *Landeskunde des Reichslandes Elsass-Lothringen* (Breslau: Königlische Universitäts- und Verlagsbuchhandlung, 1907). Photo. et coll. BNU Strasbourg.

Figure 4.8 Depictions of traditional housing styles in Alsace-Lorraine. Reproduced from E. Rudolph, *Landeskunde des Reichslandes Elsass-Lothringen* (Breslau: Königliche Universitäts- und Verlagsbuchhandlung, 1907). Photo. et coll. BNU Strasbourg.

styles were a favorite topic of research for ethnographers and anthropologists working in Alsace. These experts frequently interpreted Alsatian housing styles as visual proof of the region's link to German culture.[58]

Geography classes thus encouraged Alsatian students to adopt an ethnographer's worldview and to acknowledge their place within fixed categories of ethnic and cultural identity. Mass-produced visual archetypes provided the means to communicate these categories. Schools trained students to see representations of their regional identity in their everyday life: in the clothes their family wore, in the tools that their father used to farm the fields, in the carved wooden designs on the backs of their dining-room chairs, and in the style of roof under which they slept. Banal, everyday objects that had formerly lacked symbolic cultural value now had great importance as outer signs of Alsatian identity.

It is important to note, however, that the turn of the century's formulaic Alsatian "local image" could also act against the idea of Alsace as part of the "family of German regions": artists in civil society could subvert official classroom uses of regional archetypes. According to François Pétry, the simplistic and utopian images of Alsace-Lorraine developed during the period of the German annexation also provided the local population with a secure regional identity amid a period of transitional national identity.[59] The profusion of regional imagery did not necessarily mean that Alsace had established its place within the mosaic of German regions. It could also be viewed as a sign that Alsatian identity was flourishing outside German national identity, becoming its own "microcosm." This argument certainly applies to the work of the Francophile Alsatian cartoonist Hansi, whose widely disseminated images of Alsatian villages and the Alsatian countryside promoted a Manichean worldview of good Alsatians and evil Germans.[60]

The Alsatian "local image" was thus open to manipulation from groups who either supported or rejected the German imperial presence. Alsatian government officials, for example, used the Alsatian "local image" to create a coherent visual narrative of regional identity for the Berlin International Tourist and Commerce Exhibition in 1911 (see fig. 4.9).[61] Organized at a moment when the German Empire was experiencing economic growth, the exhibition was to provide viewers with an "educational and inspirational image" of travel and commerce in the German lands. Members of the political elite from Colmar, Metz, and Strasbourg gathered together to choose a group of visual images that they believed best represented their region. The center room of the exhibition, dedicated to Lower Alsace, was a converted farmhouse room, with wooden chairs carved by the Alsatian artist Charles Spindler, and a woman wearing a costume borrowed from the Alsatian Museum. For Lorraine, the exhibit included images of castles and memorials from the war of 1870–71, as well as a woman in the traditional white Lorrainer bonnet. The Upper Alsace section featured photographs of the district's capital city of Colmar, plastic reproductions of Kaysersberg, medieval ruined castles on mountaintops, and a panoramic view of a dairy with a painted landscape in the background. On the eve of World War I, therefore, the visual narrative for Alsatian regional identity was well established, in both imperial and regionalist circles. Its fundamental visual elements would remain intact as Alsace-Lorraine transitioned to French rule, though the French would attempt to infuse the gentle pastoral images with

Abteilung Ober-Elsaß.

Figure 4.9 Public exhibitions distilled geographic spaces like Upper Alsace into a series of simplistic images. Photograph from the official catalog of the Internationalen Fremden-und-Reise-Verkehrs-Ausstellung, 1912. Archives départementales du Bas-Rhin, 87 AL 3262.

republican political ideals. Before we can examine the political uses of regional imagery in postwar Alsace-Lorraine, however, we must first understand the key transitions in French political culture from 1871 to 1918, and their influence on images of the nation.

VISUALIZING NATIONAL TERRITORY IN THIRD REPUBLIC FRANCE

In the wake of the Paris Commune and the loss of the Franco-Prussian War, France's fragile republican government, together with associations in civil society, focused on strengthening France's internal cultural and social cohesion.[62] Until recently, historians attributed the Third Republic's achievement of stability and prosperity to its elimination of regional particularity. According to Eugen Weber's influential study, the post-1871 French state successfully transformed "peasants into Frenchmen" thanks to measures in cultural colonialism executed by Parisian government officials and schoolteachers in the provinces: banning the use of regional dialects in schools and replacing images of the Virgin Mary with the secular republican idol of Marianne.[63] More recently, however, historians such as Jean-François Chanet, Anne-Marie Thiesse, Caroline Ford, and Patrick Young have argued that the preservation of local and regional cultures was in fact a core value of French republicans.[64] In other words, conservatives such as Maurice Barrès, who believed that French identity was "rooted in the soil," were not the only ones searching for a grassroots connection to the national idea. As long as French provincials remained open to republican values and democratic political principles, Third Republic leaders believed that the region (*petite patrie*) could serve as a useful conduit for strengthening the population's commitment to the French nation-state (*grande patrie*).

This republican vision of France as the successful sum of its regional parts received enthusiastic support from the "grandfather" of modern French geography, Paul Vidal de la Blache. From the beginning of his career in the 1870s until his death in 1918, Vidal de la Blache argued that France's distinct personality lay in its ability to unify a diverse combination of regions. Compared to earlier French concepts of territoriality, according to which rulers and intellectuals thought of Frenchness as something precious and cultivated that emanated outward from the Palace of Versailles to a group of underdeveloped provinces, Vidalian geography offered a fundamentally new model for defining French identity.[65] Vidal argued that it was France's periphery that was the true incubator of Frenchness, not its center. Each province, including the recently martyred provinces of Alsace and Lorraine, contributed something unique and essential to the French nation. Vidal's belief in the importance of France's regional diversity formed the basis for his geographic methods. His famous geographic tableau—the culmination of decades-long research—gave the soil, climate, history, and industry of each part of France its own individual attention.[66] During the late nineteenth and early twentieth centuries, Vidal's idea of France's *genius loci*, its local genius, had a powerful influence on how French artists pictured their homeland's local and regional spaces.[67]

The French Third Republic's embrace of its regional diversity began with its cultivation of new female allegorical figures to represent land. In French political tradition, the unitary Republic was represented by the image of Marianne, a young peasant girl with a Phrygian cap atop her head. During the upheaval of the French Revolution and the overthrowing of the Old Regime, the female figure served as an allegory for virtue, unity, and stability.[68] She became the visual embodiment of the abstract idea of the nation as a community of people with shared political beliefs and values.[69] Some of the first popular images of the "lost provinces" to appear in France following the Franco-Prussian War pictured Marianne comforting the allegorical figures of "Alsace" and "Lorraine," personified as women in regional folk dress. Alsatian artist Adolphe Braun, for example, created a series of celebrated studio photographs with female models in Alsatian and Lorrainer folk costume wearing expressions of grief and defiance on their faces.[70] In these images, the women became visual signs that stood for the recently annexed lands. Just as the body of the French king had once been synonymous with the Kingdom of France, the women in the photographs *were* the vulnerable and victimized border territories.

Over the course of the turn of the century, however, more realistic images of French land would come to rival the female allegorical figures. As the notion of the *petite patrie* grew increasingly important for Third Republic nation builders, the French state turned to local images (*images locales*) to communicate its idealized vision of the French nation to a broad public audience. It is not surprising, therefore, that turn-of-the-century French classroom maps and geography texts looked quite similar to their German counterparts. Like their fellow German instructors, French geography teachers wanted to forge a connection between a child's familiar and tangible environment and the abstract space of the nation. To understand how teachers communicated this region-friendly vision of France to their students, we turn to the geography text that was far and away the most influential and iconic portrayal of French national territory in Third Republic schools: *Le tour de la France par deux enfants*.

A middle-level classroom geography manual that included nineteen maps and 212 illustrations, *Le tour de la France par deux enfants*, originally published in 1877 and reprinted well into the twentieth century, was one of the best-selling French books of all time. Authored by a woman named Augustine Fouillée, who disguised her gender with the pen name G. Bruno, the book told the story of two orphan boys from Lorraine, brothers André, age fourteen, and Julien, age seven, who escaped from German territory in search of their French homeland. A travelogue told from the perspective of the two boys, the text's description of French territory skillfully interwove the core values and principles of Third Republic political culture with visual images of France's most significant regional landmarks (see the boys' route across France in fig. 4.10).[71] Though her geographic descriptions focused on those regions that were formally part of French territory after 1871, Fouillée's decision to base her narrative on the travel experiences of two boys from annexed Lorraine ensured that the memory of the lost eastern borderland—and the French boys and girls left behind there—would remain in the French public consciousness until the First World War.

The idea of a "Tour de France" was well embedded in French culture. Since the Old Regime, journeymen workers (*compagnons*) had traveled from one French city to the next, often for years at a time, to perfect their craftsmanship at different regional workshops. The young journeyman's travels were akin to a rite of passage: upon completion of his Tour de France, the journeyman was ready for membership in a guild. Augustine Fouillée's

Figure 4.10 Augustine Fouillée's story of André and Julien, young boys from annexed Lorraine who journey across France to learn about its history and culture, became a foundational geography text for French school-children. Image from G. Bruno, *Le tour de la France par deux enfants* (Paris: Belin, 1905). Photo. et coll. BNU Strasbourg.

classroom reader represented a late nineteenth-century interpretation of the journeyman's well-trodden Tour de France. Instead of a craftsman's rite of passage, her text emphasized the civic virtues and collective morals that formed an educational rite of passage that was necessary for children to become good French citizens. For Fouillée, France was more than a geographic area: it was a great nation with a history of glorious events and leaders. Her book's popularity lay in the successful connection that it forged between national memories and the visible, tangible geography of France. "If the memory of *Le tour de France* is more topographic than historical," write Jacques and Mona Ozouf, "surely it is because Mme Fouillée, good teacher that she was, understood that collective memory depends more on places than on dates."[72]

In the preface to her book, Fouillée emphasized the importance of a child's experiential understanding of French territory. Like the German geography teachers of the time, Fouillée structured her text around the principle that empirical observation and sensory experiences provided the best foundation for knowledge:

> Our schoolmasters know how hard it is to give the child a clear idea of the homeland [*patrie*], or even simply an idea of its territory and resources. For the schoolchild, the homeland only represents something abstract, from which, more often than we believe, he can remain estranged for a long period of his life. To inspire his mind [*ésprit*], the homeland must be rendered visible and alive. With this goal, we have tried to benefit from the interest that children take in travel tales. In telling them about the courageous voyage of two young Lorrainer boys across the entirety of France, we wanted them to see and touch as well.[73]

Fouillée's goal of rendering the homeland "visible and alive" (*visible et vivante*) mirrored the concept of a "homeland image filled with life" (*lebensvolles Bild der Heimat*) found in German geography texts from the same period. The difference from the German texts, of course, was that the Lorrainer boys' tour of France was not written from the perspective of provincial authors with a vast store of local knowledge. Published in Paris, Fouillée's text provided a geographical overview of provincial France that met the approval of urban editors living in the nation's capital city. Julien and André were caricatures of what a French woman in the 1870s believed Alsatians and Lorrainers to be thinking and feeling.

Who were André and Julien, the characters that Augustine Fouillée invented? They grew up in the Lorrainer village of Phalsbourg, a historic fortress town that the Marquis de Vauban had built in the seventeenth century. During the Franco-Prussian War, the Germans laid siege to the city; the boys' father, Michel, suffered a serious leg injury trying to defend the town from the invasion. While working as a carpenter after the war, Michel's unstable leg gave way and he fell from the scaffolding at his construction site. Before his accident, Michel had already made plans to leave annexed Lorraine for France and was working to collect enough money for the trip. On his deathbed after his terrible fall, Michel turned to his sons and told them that they must take up his dream of escaping German territory for France. "His eyes turned toward the open window where a corner of big blue sky was showing," wrote Fouillée with her characteristic drama, "over there, he seemed to search for the horizon of the dear homeland's distant frontier that he would not reach, but where his two sons, without help from now on, had promised him they would go" (10). Despite their grief, the two boys, particularly the eldest, André, exhibited strength and courage, vowing to fulfill their promise to their father. In reality, of course, the boys did not have to flee Alsace-Lorraine to remain French; residents of the border territories in fact had several years to opt for French rather than German citizenship. But a calm and peaceful emigration from Alsace-Lorraine was not what Fouillée had in mind for her two heroes.

In planning their dangerous escape from German Lorraine, André and Julien relied on maps, introducing the reader early on to the social utility of cartography. With the help of a forest guard who was a friend of their deceased father, André procured a large map of his department, "one of the pretty maps drawn by the General Staff of the French Army, where even the smallest paths were indicated" (17). Their father's friend pointed out the nearby routes, paths, and shortcuts that would help the boys cross the nearby frontier at the crest of the Vosges, a frontier vigilantly defended by armed German guards. After studying the departmental map and its various paths, André drew his own map, which indicated the route that he and his brother would take to the German frontier, punctuated by selected landmarks: "Here, he wrote, a fountain; there, a group of beech trees through the firs; further, a torrent with a ford to cross it; a rock peak that twists the path; a tower in ruins" (17). Thus, in the first section of her travelogue, Fouillée suggested to the reader that knowing how to use a map demonstrated a resourcefulness and a can-do spirit that defined a citizen of the French republic. "What good would it serve me," remarked Julien, "to be the best student in the School of Phalsbourg up until the age of thirteen if I cannot succeed in guiding myself with the aid of a map?" (17). On the chosen night, the self-reliant brothers set out for the French-German border crossing, their little travel sacks tied to the ends of sticks resting on their small shoulders. Overcoming several obstacles, the boys reached the border. Just as they were crossing, the sun rose over the horizon; it was dawn, and the boys were in France, prepared to embark on their adventure. "Beloved France," they cried, "we are your sons, and we want to remain worthy of you for the rest of our lives!" (25).

In the several hundred pages that follow, André and Julien traveled from region to region in France, learning about their great nation through their direct experiences with its natural and built environment. "It is the physical contact with the land," write Jacques and Mona Ozouf, "that gives the journey its dynamic intensity."[74] *Le tour de la France* presented the relationship between the French regions as one of harmonious cohabitation, with each regional element contributing to the richness of the French nation. The purpose of the text's illustrations—maps and pictures engraved especially for the book—was to add a documentary, encyclopedic quality to the boys' encounter with French territory: the regional monuments and landmarks that they saw and the inhabitants that they met.[75] Together, the images formed an iconographic program, tying the study of French geography to a series of canonized regional images similar to those used in Alsatian textbooks under German rule. The text's image of the fir tree from Lorraine, for example, would become so engrained in the French collective memory that it would reemerge in numerous postwar French pictorial representations of Alsace-Lorraine.[76] Moreover, the mental map of France promoted in *Le tour de la France*—a hexagon still intact—reflected a biased stance on French border politics. By incorporating the annexed borderland into her visual narrative of French national territory, Fouillée reinforced the cultural justification for revanche: the declaration of war to reclaim Alsace-Lorraine. Her classroom reader was, of course, not alone in this regard. Most classroom wall maps in turn-of-the-century France continued to include the lost provinces in their cartographic representations of French national territory; Alsace-Lorraine was simply colored purple, to signify mourning, or white, to signify absence.

Thus, while Paul Vidal de la Blache played an important role in popularizing a version of French territoriality that valued regional diversity, it was a little-known woman, writing under a pseudonym, who created the most influential mental map of Third Republic France. How do we make sense of this paradox? Bernard Lightman has argued that the field of popular science offered nineteenth-century women an opportunity to make contributions outside the more restricted world of professional science.[77] The same can be said for women in the fields of cartography and geography. Given their lack of access to university posts and cartography positions within the army, it is perhaps not surprising that women turned their efforts toward the popularization of geography through education. In fact, evidence from across the globe reveals an interesting pattern: the most popular classroom geography books in nineteenth- and early twentieth-century France, Sweden, and the United States were all authored by women.[78] After World War I, of course, Augustine Fouillée and her fellow experts in geographic pedagogy would be faced with new challenges. Alsace and Lorraine were French territories again and needed to be remapped.

THE *HEIMAT* BECOMES THE LITTLE REPUBLIC

In November 1918, the Parisian government transported 1,227 busts of Marianne to Alsace-Lorraine. Together the Mariannes weighed forty-seven tons and covered an area of four hundred cubic meters.[79] The French state provided local officials in Alsace-Lorraine with stencils of the republican coat of arms and told them to remove statues of German imperial eagles from public places, taking care not to pull down Napoleonic eagles by accident.[80] In the early months following the armistice, the Parisian government viewed the republican coats of arms, emblazoned with the words "Liberty, Equality, Brotherhood," and Marianne's Jacobin personification of France as important symbolic elements in their repossession of Alsace-Lorraine. In the years following, however, French institutions in Alsace-

Lorraine also seized upon local imagery as a vehicle to educate Alsatian schoolchildren about the French nation. Many of the images were familiar from the German period: idyllic village scenes, half-timbered houses, storks, and people in regional dress. The French, however, framed these images of local life according to their particular understanding of a region's role in national political culture. French classrooms used the Alsatian *image locale* to reinforce the concept of a Republic united in its regional diversity. The French image of the Alsatian village *clocher*, while offering some parallels to the German Empire's concept of *Heimat*, was firmly rooted in French republican tradition and lacked any promise of a decentralized or federalized political order.

One of the first geography texts printed in the newly French Alsace-Lorraine focused on local topography and imagery in a manner reminiscent of previous German texts.[81] Its first lesson, entitled "Little Charles Is Lost," revolved around the image of the *clocher*, the village bell tower. The story was about a boy named Charles, and it went as follows: One day, Charles left his home to see a neighboring village, crossing a "beautiful forest of fir trees" to get there (reminiscent of the imagery from *Le tour de la France*). Charles became so distracted by the animals and plants that he saw in the forest that he got lost. But Charles was able to save himself using the skills that he had acquired in his geography class. He looked up to the sun, calculated his directional orientation, and realized that he had traveled too far south of his village. After several minutes of walking, Charles saw the *clocher* of his village, the visual signal that he was home. The geography text thus highlighted the church tower as a child's primary point of orientation, a reliable directional compass for children when they moved about and worked in their village's surrounding landscape. But it also represented something greater: the church tower also played the role of a moral compass. In modern France, the church tower was a powerful visual symbol for local pride and identity.

In his study on village bells, Alain Corbin discusses the village *clocher* as a core spatial and visual concept in French culture. According to Corbin, Third Republic politicians sought to appropriate church bell towers from village priests as a means of symbolically possessing and secularizing rural localities' most prominent visual marker.[82] Once they desacralized the church tower, republicans eagerly manipulated its image to their advantage. Their visual strategy was to disconnect the form of the church tower from its function; they transformed the empty shell of the church tower, stripped of any religious meaning, into a secular image of France as a union of thousands of identical communities. This political manipulation of the church tower's image was not unlike that of the German *Heimatler*, whose engravings portrayed German churches as ambiguously Catholic or Protestant, hoping the lack of attention to denominational detail would symbolically heal the religious rift created by Berlin's *Kulturkampf* (cultural war) on the empire's Catholic provinces.[83] The republicans' symbolic fusing of the locality and the Republic through the quaint image of the bell tower helped to undo the urban stereotype of villages as antimodern spaces characterized by their insularity and parochialism, or in French, their *esprit de clocher*.[84]

Tying geography to civics instruction, teachers in postwar Alsace merged the image of the secularized *clocher natal* with the idea of the Republic. Local topography became a metaphor for France's central government. One text explained: "Our commune, with its public, forms a republic. The village is its capital. In this capital lives the president (the mayor) and his council and its employees. The country that surrounds the capital contains verdant fields, streams, hills, a hamlet, and several isolated houses."[85] French schoolteachers therefore transformed the previous German vision of the Alsatian village as one of the empire's numerous *Heimats*, into a French vision of the Alsatian village as an ideal political community run like a miniature republic. The text's use of the village as a microcosm of the national community demonstrated how dramatically French nation-building strategies had changed since 1870, when Alsace and Lorraine had last been part of French territory. Teachers now taught French patriotism from the vantage point of local and regional realities. The text's message was: culturally and geographically, French people were diverse, but as citizens, they formed a unified whole. Their shared belief in republican democracy was the core value that bound them together.

As a general rule, moreover, Alsatian geography instruction fitted into a larger pattern of postwar French educational policy in which local schoolteachers could choose their own materials and design their own lesson plans.[86] Take, for example, the hand-drawn map images that one local schoolteacher, Fr. E. Hennigé, used for geography instruction in his tiny community of Fréland—a village of only 324 houses—which offered many parallels to the spatial concepts that schoolteachers employed under German rule (figs. 4.11–4.14).[87] Relying on his personal knowledge of local geography, Hennigé described Fréland's unique physical surroundings, noting the relatively strong wind that blew through its valley compared to the nearby village of Bonhomme, which was sheltered from the wind. He noted the special composition of Fréland's rocks, also comparing them to those of Bonhomme. The text also taught students about Fréland's local history and folklore: the village was named after a colony of coal miners who did not charge any taxes and therefore called their town Freiland, German for "free land." Situating Fréland within the greater Alsatian region, Hennigé explained that Strasbourg was the region's "commercial center" and Mulhouse was its "worker city." Even under the centralized rule of the French republic, therefore, Alsatian geography instruction continued to emphasize

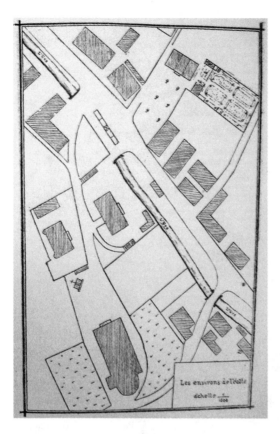

Figure 4.11 When Alsatian and Lorrainer schools transitioned to French rule after 1918, geography curriculum continued to emphasize the significance of local spaces. Geography lessons began with a map of the school and its surrounding streets. Fr. E. Hennigé, *L'Alsace: Géographie locale et régionale à l'usage des écoles primaires des lycées et des collèges avec modèle d'une monographie géographique du lieu de domicile, carte en couleurs, plans et croquis* (Colmar: Société alsacienne d'édition Alsatia, 1920). Photo. et coll. BNU Strasbourg.

the hometown as the primary spatial unit that underpinned the larger territories of the region and the nation.

Working in tandem with maps, pictorial images established a powerful visual narrative of the interconnections among locality, region, and the nation in Alsace-Lorraine after World War I. Canonized regional images formed the basis for Christian Pfister's *Alsatian Lectures*, one of the most important history and geography texts in postwar Alsatian schools. Pfister, an Alsatian medievalist who became rector of the University of Strasbourg during the 1920s, played a critical role in the region's reintegration into France. A classic *image locale* archetype, the cover of the text assembled a formulaic collection of regional images that could never have existed together in real life. Fir trees, grapevines, storks' nests, and the Cathedral of Strasbourg were grouped together in such an elegant and inviting fashion that it was difficult for the viewer to notice the unnatural fusing of urban and rural scenes. The book's text likewise promoted a "regionalized" view of Alsatian villages as communities that were all the same. Quoting from Alphonse Daudet, the celebrated turn-of-the-century French writer, the text reads: "What are the names of these pretty Alsatian villages that we encountered on the side of the roads? I can't remember any names right now, but they all resembled

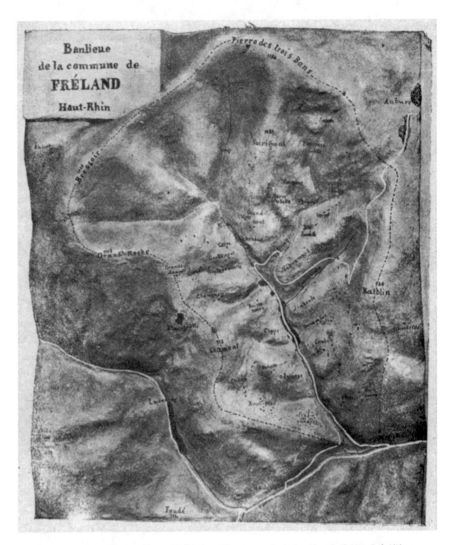

Figure 4.12 Map of the Alsatian village of Fréland used for geography instruction. Fr. E. Hennigé, *L'Alsace: Géographie locale et régionale à l'usage des écoles primaires des lycées et des collèges avec modèle d'une monographie géographique du lieu de domicile, carte en couleurs, plans et croquis* (Colmar: Société alsacienne d'édition Alsatia, 1920). Photo. et coll. BNU Strasbourg.

each other so much, especially in the Haut-Rhin, that after having traveled very far for hours, it seemed to me that I had only seen one village: the main street, the little windows framed in lead, covered with roses, open doors where the old people smoked their pipes."[88] French schools thus presented Alsatian children with a rubber-stamp image of their region, just as the German schools had before the war. Alsace's villages appeared as "one village." This time, of course, schoolteachers tied their regional archetypes to the mental image of France as a "family of regions," an image already popularized by turn-of-the-century regional tourism and classroom materials such as Fouillée's *Le tour de la France.*

x

Comment appelle-t-on une nouvelle envoyée par le télégraphe ? De quelle manière les dépêches arrivent-elles jusqu'à notre bureau de poste ? Quel est le jour de marché de Kaysersberg ? — de Colmar ? Où se rendent ordinairement les touristes qui passent par le village ?

La banlieue de Fréland.

La banlieue de Fréland a une superficie de 1.937 hectares, qui se partagent en 950 hectares de forêts, 550 hectares de prés et de pâturages, 400 hectares de champs et le reste en terrain vague (maisons, chemins, rivières, terre inculte).

La banlieue de Fréland comprend *tout le bassin de l'Ure.* Les limites passent sur la crête des montagnes qui entourent le vallon. Elle est bornée au nord-est par Aubure, au nord par Sainte-Marie-au-Mines et Lapoutroye, à l'ouest et au sud par Lapoutroye, à l'est par la banlieue de Kaysersberg qui renferme la forêt de Sigolsheim.

Voir la borne qui se trouve sous le petit pommier, dans le pré à gauche de la route qui descend à la gare, et deux autres dans les prés entre la route et la rivière. Une vieille croix (de l'année 1719) marque la limite sur la route de Hachimette. (Au milieu du village, il y a une autre croix de l'année 1718.)

Expliquez : Pierre des Trois Bans. Montrez la direction dans laquelle se trouvent les communes voisines. Quelle est la commune qui est tout près de Fréland, sans en toucher la banlieue ? Dans quelle direction est-elle située ?

Les environs de Fréland.

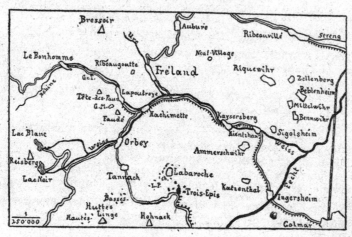

Figure 4.13 Map of the villages surrounding Fréland. Fr. E. Hennigé, *L'Alsace: Géographie locale et régionale à l'usage des écoles primaires des lycées et des collèges avec modèle d'une monographie géographique du lieu de domicile, carte en couleurs, plans et croquis* (Colmar: Société alsacienne d'édition Alsatia, 1920). Photo. et coll. BNU Strasbourg.

Figure 4.14 Through a series of spatial expansions, Hennigé's geography text connected Fréland and its network of villages to the larger national territory of France. Fr. E. Hennigé, *L'Alsace: Géographie locale et régionale à l'usage des écoles primaires des lycées et des collèges avec modèle d'une monographie géographique du lieu de domicile, carte en couleurs, plans et croquis* (Colmar: Société alsacienne d'édition Alsatia, 1920). Photo. et coll. BNU Strasbourg.

Despite the strong continuity between pre- and postwar visual representations of Alsatian territory, some French interest groups challenged the traditional look of Alsatian local images. New kinds of visual images, representing Alsace as a dynamic zone of industrial economic growth, began to circulate in the popular press. The Strasbourg Tourist Commission's decorative wall map, *Beautés de l'Alsace: La nature, les monuments*, drawn by the celebrated Alsatian graphic artist Balthasar Haug and published in 1932, underscored the connection between visual propaganda and France's elaborate postwar schemes for Alsatian trade and industry. The map's function was purely iconographic: it did not contain any directional indications or longitudinal and latitudinal markings. Its sole function was to promote a certain view of Alsace as a region rich in both picturesque sites and industrial potential. Its visual style symbolically "beautified" the presence of industry and resource extraction facilities on the Alsatian landscape.

In terms of its iconographic language, the map's symbols for an "artistic town" (*ville d'art*) and an "industrial town" (*ville d'industrie*) resembled each other quite closely. The industrial town's factory buildings had the same roofs as the medieval houses in the artistic town, and its smokestack towers appeared harmonious on the skyline. There was a political and commercial purpose behind this iconographic program. Following World War I, economic planners in Paris, together with Alsatian entrepreneurs, undertook a number of large and costly industrial projects, including the Great Alsatian Canal with its eight hydroelectric stations and a new port of Strasbourg

Figure 4.15 The cover of Christian Pfister's book was a typical *image locale*, assembling a formulaic collection of regional images including storks, half-timbered houses, grapevines, and the Cathedral of Strasbourg. Christian Pfister, *Lectures alsaciennes: Géographie, histoire, biographies* (Paris: Librairie Armand Colin, 1919). Photo. et coll. BNU Strasbourg.

Figure 4.16 A tourist map from the interwar years incorporated "artistic" and "industrial" towns into a visual display of the "Beauties of Alsace." Balthasar Haug, *Beautés de l'Alsace: La nature, les monuments* (Strasbourg: Commission du Grand Tourisme, 1932). Archives départmentales du Bas-Rhin.

linked to oil refineries and potash mines (see the map key's symbols for *canal existant/projeté*, *pétrole*, and *potasse*). Though large parts of these projects were still in the blueprint stage when the map was published in 1932, the image gives us insight into the government and private sector's desire to transform Alsace into a thriving zone of energy production and economic exchange on the Franco-German border.[89] The map created a symbolic landscape in which modern oil refineries and factories blended seamlessly with traditional regional landmarks like castles, abbeys, and winemaking villages. It was a vision for Alsace's future, however, that would not come to fruition until France and Germany made peace after World War II.

Persuading ordinary people in Alsace-Lorraine to feel a sense of loyalty toward territories that were too large for them to see or experience in their entirety was one of the greatest challenges facing French and German nationalists. The makers of survey maps and national atlases attempted to solve this problem by creating imperial images that pictured the borderland as an appendage of a larger nation-state or empire. But these types of images, scientific and cerebral in their form of visual communication, could not easily connect with the senses or emotions of ordinary provincials. Some nationalists chose instead to create maps designed to forge a psychological connection between a citizen's local community—an intimate, beloved, and familiar place—and the abstract and faraway territorial entities of "France" and "Germany." They developed innovative cartographic techniques that flipped the direction of their map's spatial orientation from the capital city to the provincial village, encouraging citizens to visualize their national territory as a space that flowed naturally outward from their hometown. Blurring the lines between experienced and abstracted (paper) space, the maps trained people to see their national identity in their local surroundings: their village streets, bell towers, fields, and classrooms. The danger, of course, was that maps focusing on such small spaces could be appropriated by homegrown regionalists who saw Alsace's picturesque villages as inherently Alsatian, rather than French or German, cultural spaces. Nationalists therefore had to develop techniques of visual training and pedagogy, both inside and outside classrooms, to ensure that images of local and regional space were always tied to a universal image of the nation.

Like the informal network of cultural boundaries that spread across thematic maps of Europe beginning in the nineteenth century, maps of local and regional space held no legal or jurisdictional authority. Their power lay in their ability to manipulate public opinion and perform the work of nation and region building. In the following chapter, we turn to another body of unofficial, but highly persuasive, territorial images that were created by and for European citizens. These images interpreted one of the most spectacular natural environments in the French-German borderland: the Vosges Mountains. In exploring how nationalists and regionalists transformed the rugged mountain range into a cultural landscape, we discover how nature itself became a socially produced space in modern Europe.

CHAPTER 5

Maps for Movement

Extending hundreds of kilometers across northeastern France, the Vosges Mountains mark the end of the Parisian geological basin and the beginning of Alsatian territory. The High Vosges, culminating in the majestic Ballon d'Alsace, are composed of crystalline granite while the Low Vosges are made of a red and pink limestone coveted by Alsatian architects for its malleability and colorful glow. The mountains are covered with a seemingly endless expanse of forests filled with firs, pines, spruces, and beeches. Numerous lakes and waterfalls, left over from the glaciers that at one time carved out the massive pieces of stone, stretch across the wooded terrain. Unlike the young Alps with their sharp peaks, the plates beneath the old Vosges are slowly collapsing downward. From a distance, the crests of the Vosges appear as a series of rounded, dark green curves, gently rising and falling across the horizon. On a clear day, the silhouette of the mountains against the sky produces a dazzling color that Goethe celebrated in his poetry and French nationalists once referred to as "the blue line of the Vosges."[1]

Besides their natural beauty, the Vosges are renowned for their physical reminders of the lost civilizations that once inhabited the border area between modern-day France and Germany. If a traveler looks up at the mountains from the Alsatian Plain, ruins of fortified castles become visible on several of the crests. In medieval times, when Alsace was still firmly within the German sphere of influence, feudal lords built the castles to defend their lands against warriors sweeping across the plain below. A total of 121 castles are scattered throughout the Vosges, transforming the mountainous environment into a historic human landscape. Deeper within the Vosges are still older vestiges of the Gallo-Roman and Celtic civilizations that once inhabited the area. These ancient cultures believed that certain peaks emitted supernatural vibrations, and they built sacrificial temples on several remote mountaintops to worship their pagan gods. Archaeologists from modern France and Germany have revisited these mysterious sites, excavating the soil for evidence of their distant ancestors.

When the French-German border moved westward from the Rhine to the Vosges after the Treaty of Frankfurt in 1871, the mountains attracted the attention of civilians who were eager to develop patriotic maps of the new border site. The hiking maps of the Vosges that citizen surveyors designed, printed, and circulated offered an alternative perspective on the borderland's geography to the perspectives of military, administrative, or linguistic maps. Their main purpose was to serve as navigational tools for tourists who wanted to experience the mountainous environment firsthand. The group of citizens that invented the first-ever set of hiking maps for the Vosges, during the early 1870s, was the Vogesenclub, one of the most powerful outdoor associations in German Alsace-Lorraine, which later became the French Club Vosgien after World War I. Both the German and the French versions of the club sold their cartographic images of the Vosges on the commercial market, where they found a broad audience of map readers eager to learn more about the cultural and historical topography of the mountains.[2] Foldable and cheap, designed to be sweated on and dirtied, the clubs' hiking maps encouraged French and German citizens to define and experience their national boundary through physical movement.

The practice of marking territorial boundaries with one's feet was not new to Europe. French monarchs once defined the geographic limits of their kingdom by making their bodies visible to their subjects through ceremonial visits to far-off provinces.[3] For generations, European farmers perambulated around their rural parishes in order to commit their communal boundaries to memory.[4] But the border tourism that took place in the disputed Vosges Mountains during the late nineteenth and early twentieth centuries was fundamentally different from these earlier forms of walking itineraries. The number of people who wanted to see and experience the French-German border firsthand increased exponentially thanks to the rise of mass tourism: the invention of railroads and the expansion of a European middle class with disposable income brought never-before-seen numbers of travelers to the Vosges. Furthermore, as Europe's territorial boundaries took on new importance as national symbols, tourists began to think of border viewing as an act of patriotism accompanied by the singing of national hymns, visits to historical monuments, and ceremonies for their nation's war dead. Beginning in the late nineteenth century, civil associations like the Vogesenclub and the Club Vosgien took advantage of the newfound popularity of border tourism, selling maps, panoramas, and guidebooks that promoted a certain idea of the Vosges Mountains as a quintessentially French or German landscape.

Landscape is one of those unofficial territorial terms—like homeland, *Heimat*, or *patrie*—that has no formal legal status but has nonetheless served as a highly influential spatial concept in modern European culture. From an etymological standpoint, there is a clear connection between the French word for country, *pays*, and a subjective landscape image of that country, *paysage*, a connection echoed in the relationship between the German words *Land* (land or country) and *Landschaft* (landscape). "National identity," argues Simon Schama, "would lose much of its ferocious enchantment without the mystique of a particular landscape tradition: its topography mapped, elaborated, and enriched as homeland."[5] More recently, in his study on water and the making of modern Germany, David Blackbourn likens the concept of landscape to a "mental topography" in which the individual mind enriches the natural environment that it sees with its own particular set of preconceptions, cultural values, and political orientations.[6] Charles Harrison has emphasized the relationship between landscapes and "collective viewing," arguing that the creation of national landscapes, particularly in border areas, depends on the collective visual training of large groups of people.[7] While each of these authors offer important insights into the social production of landscapes in modern Europe, evidence from Alsace-Lorraine's map archive suggests that we need a far less static definition of the term "landscape" than these authors have proposed.

When scholars have analyzed the historical development of European landscapes, they have typically focused on pictorial images such as lithographs, photographs, and paintings. This chapter does not dispute the fact that each of these types of visual images is critical to understanding the changing role of landscapes in European political culture. It argues, rather, that our analysis of historical landscapes has suffered from the exclusion of cartographic sources. Maps were fully part of the visual culture of their time; they interacted with, and gave meaning to, the photographs and lithographs that appeared next to them in mass print. Hiking maps demonstrate how the freeze-framed views presented in pictorial landscape images were in fact linked together through topographical itineraries that demanded physical movement. As Tim Ingold has recently argued in his study on landscape, our perception of our surroundings comes not only through our eyes, but also through our entire bodies and through all our senses.[8] French and German hikers in the disputed Vosges Mountains developed an intimate understanding of their national landscapes through the maps that guided their feet across the ground. As a cartographic genre, hiking maps helped European citizens to identify with their nation by encouraging their physical contact with border terrain.[9]

Hiking maps and panoramas, moreover, shed new light on the physical changes that the French and Germans imposed upon their border territories during the nineteenth and twentieth centuries. Just as the Gardens of

Versailles provided a powerful setting for French kings to demonstrate their absolutist political philosophy through the shaping and ordering of plants, we can interpret hiking clubs' beautification and trail-making initiatives in the Vosges Mountains as three-dimensional expressions of French and German political programs.[10] Over the course of the nineteenth and twentieth centuries, the French and Germans acted as gardeners of the Vosges, hammering in park benches, erecting viewing towers, propping up collapsing castles, and establishing war memorials on shell-blasted hills. French and German hiking maps, in other words, did not simply reflect a static natural environment in the Vosges Mountains. They reflected a territory in constant change thanks to the landscaping initiatives of patriotic European citizens.[11]

FRENCH STEPS TOWARD NATIONALIZING
THE VOSGES LANDSCAPE

The first modern travelers to the Vosges intent on mapping human values and emotions onto geography were the bourgeois travelers of the late eighteenth and early nineteenth centuries. Unlike later tourists and hikers inspired by nationalist and regionalist movements, these travelers were an eclectic international group. They included Frenchmen such as Victor Hugo, Germans such as Johann Wolfgang von Goethe, and Alsatian notables such as the organ maker Jean-André Silbermann and the Protestant minister Jean-Frédéric Oberlin. With no access to hiking maps or trails, they had to pay local guides to lead them through the remote crests and valleys of the Vosges on horseback, in carriages, or by foot.[12] Along the way, the men wrote poetry, obtained plant samples, and drew sketches of beautiful scenery. Many of their tours of the Vosges were typical of the era of the "picturesque and romantic voyage," a form of landscape tourism popular among the European elite. "The picturesque voyage," Alain Corbin explains, "is composed of a series of 'points of view' determined by the choice of successive stations at the sites chosen for their capacity to frame an image."[13] These eighteenth- and early nineteenth-century "viewing stations" offered meditative scenes and portraits of the sublime that fulfilled a generation's fascination with human emotions. Furthermore, the viewing stations provided a worthy educational experience for young men and women: the incorporation of ancient monuments and castles into landscape scenes reflected an idealized view of provincial tourism as a form of time travel into Europe's past.

Before the age of mass tourism, this elite group of travelers to Alsace and Lorraine voyaged alone or in small groups. As members of the rising bourgeois class, the men and women were drawn to remote places such as the Vosges as a haven to escape the stresses of modern life and to cultivate a self-knowledge (translated in German as *Bildung*) through their direct physical encounter with Europe's lost civilizations. The modernizing societies from which the tourists emerged demanded hardworking, self-assured individuals to lead their people into a new age of social and economic progress. Nature offered a place for these strong leaders to reengage with their deepest emotions in the overwhelming presence of the wild. Caspar David Friedrich's iconic *Wanderer above a Sea of Fog* (1817) perfectly captures this emerging bourgeois culture with its image of a lone man, his back turned to the viewer, meditating on a sweeping German mountain range from an outcrop of rock.

Figure 5.1 Caspar David Friedrich, *Wanderer above a Sea of Fog*, 1817. Photo Credit: bpk, Berlin/ Hamburger Kunsthalle, Hamburg, Germany/ Art Resource, NY.

While early nineteenth-century explorers analogous to the man in Friedrich's painting embraced to the Vosges as their muse, they did little to ensure the preservation of the mountains' scenic and historic sites. This attitude began to change in the mid-nineteenth century, when the Alsatian and Lorrainer bourgeoisie came to the realization that the picturesque sites which had inspired romantic-period travelers were languishing in a state of severe physical degradation. Beginning under the July Monarchy, and particularly under the Second Empire, Alsatian gentlemen undertook it as their civic duty to preserve the ruined castles and picturesque scenery of the Vosges for future generations. They understood that their task was too large to accomplish as individuals: they needed to make a unified, organized, and sustained effort to accomplish their goal. Following a trend toward historical preservation across provincial France, in 1855 a group of middle-class Alsatian gentlemen founded the Society for the Conservation of Historical Monuments in Alsace. The goal of the association, whose members included the prefect of the Bas-Rhin and other administrators with close ties to Paris, was to catalog the historical and picturesque sites in the Vosges systematically as part of a wider French effort to identify and preserve certain sites on French soil as part of the "national patrimony." In addition to creating a written record, the society sponsored two well-known Alsatian artists—photographer Adolphe Braun and lithographer Jacques Rothmuller—to create a visual archive of the sites. This project received the full support of the French emperor. Braun's *Alsace Photographed* (1859), a series of large-format photographs of the landscapes and edifices in the Vosges and rural Alsace, was so admired by Napoleon III that he named Braun "Photographer of His Majesty the Emperor" and purchased five copies of his work for the Crown Libraries.[14]

Though the French imperial government provided financial support for the Alsatian historical society's work, it was the association's members themselves who took on the responsibility of discovering sites of national patrimony within Alsatian territory. The citizens who formed the association thus became the cultural equivalent of the state's professional topographical surveyors. Rather than canvassing the Alsatian terrain for geodesic or trigonometric measurements, they searched for sites of picturesque and historical interest that reflected their particular vision of Alsace as a fully French region. Though many of the castles and ruins in the Vosges dated to the Holy Roman Empire, when Alsace was within the German sphere of influence, the nineteenth-century Alsatian elites categorized the monuments as markers of "local memories," thereby bringing the arguably German sites into the French national patrimony by way of their regional character.[15] The sites that they located constituted what association members referred to as an Alsatian "open-air museum." Several members of the society published a guidebook, which they entitled *Picturesque and*

Historical Museum of Alsace, which introduced the reader to a mental topography for the region.[16] The title of the guidebook turned on its head the conventional public understanding of a museum as an enclosed building that housed a group of objects and artifacts within a secure indoor space. The concept of a sprawling outdoor museum, the association hoped, would breathe new life into the public's experience of history, helping citizens to connect with the past in their everyday lives and in their local environments. The invention of the "outdoor museum" thus marked a significant step toward the construction of a modern Alsatian landscape composed of sites of memory that served as visual embodiments of regional and national patrimony.[17]

This first attempt to popularize a French national gaze in the Vosges was, however, tempered by the Alsatian historical society's deep respect for individual autonomy. Steeped in the bourgeois mindset of self-cultivation, the society's members believed that the individual tourist must be allowed to have his or her own unique sensory experience during a visit, based on his or her unique physical perceptions. In the introduction to *Picturesque and Historical Museum of Alsace*, the three authors explain their decision to adopt first-person narratives in describing their itinerary through the Vosges and the Alsatian Plain:

> Each man, taken as an individual, sees, feels, and understands according to a moral disposition or dispositions that are entirely particular to him. . . . It is craziness to want to impose on him, in certain circumstances, a way of seeing, feeling, and understanding that most often is not and cannot be his. . . . The intelligent reader does not want us to dominate him. . . . He prefers indications to solutions. It is good to tell him what we think, but not what he must think.[18]

The guidebook paired the first-person narratives of the authors, Levrault, Mossmann, and de Morville, with the detailed etchings of Jacques Rothmuller, an artist from Colmar. The visual images provided a fitting supplement to the guidebook's texts. Lithographic printing technology enabled Rothmuller to use subtle grey tones to create the same kind of meditative, almost dreamlike image of Alsace that the written narratives had evoked.

De Morville's description of his journey to the historic Alsatian village of Ribeauvillé and its nearby castle exemplified how the guidebook's lyrical first-person accounts painted a vivid picture of the region's picturesque sites in the mind of the reader. "Seated coquettishly and nonchalantly at the foot of a rocky promontory,"[19] De Morville described Ribeauvillé in passive feminized terms that contrasted with his masculinity as an active and mobile male explorer. Upon reaching the Castle of St. Ulrich, situated on a mountaintop above Ribeauvillé, De Morville injected a political tone into his narrative. Connecting his idea of a French national community to the ancient castle's crumbling ruins, De Morville remarked that "the history of a people is written entirely in stone."[20] Rothmuller's accompanying image of the Castle of St. Ulrich established an air of mystery and intrigue around the castle (see fig. 5.2). Two miniscule figures pictured below the castle hinted at the target audience for the text: bourgeois men in top hats, accompanied by their dogs. At a cost of thirty-eight francs—more than a week's earnings for a skilled worker—such gentlemen were the only people who could afford to purchase the book.

Missing from the guidebook's rich collection of visual images, however, was a map appropriate for tourist use. While the guidebook included a section from the French Army's *Carte d'État-Major*, the map's battle-oriented graphic language appeared out of place next to a text about place, emotion, and cultural memory. Maps that aligned more closely with the pub-

Figure 5.2 Jacques Rothmuller's lithographic image of the Castle of St. Ulrich near Ribeauvillé. Image from L. Levrault, Th. de Morville, and X. Mossmann, *Musée pittoresque et historique de l'Alsace: Dessins et illustrations par J. Rothmuller* (Colmar: J. Rothmuller, Editeur, 1863). Photo. et coll. BNU Strasbourg.

lic's growing passion for regional landscapes were only just beginning to appear in the 1860s. One such publication, the French *Illustrated National Atlas*, showcased new kinds of cartographic images that focused specifically on regional memories, personalities, and natural settings (see plate 14).[21]

Though the Society for the Conservation of Historical Monuments in Alsace created an influential mental topography for the Vosges Mountains, it did not establish a monopoly over the kinds of landscape images that circulated in Alsatian civil society. While the historical association argued that Alsace was part of a distinctly French national space, the Société vogéso-rhénane (Vosges-Rhine Society), founded in 1868, emphasized the cross-border *similarities* between Alsace and Baden, the neighboring German state to the east. The visionary behind this new organization was the eminent Alsatian scientist Frédéric Kirschleger, a self-described "botanist-geographer" who published a catalog of all of the native plant species in the Vosges.[22] Instead of viewing the Rhine as a natural point of rupture between distinct territorial entities, Kirschleger emphasized the natural—and biological—relationship between the French and German lands that lay on the two sides of the river. "The land of Baden, from the topographical point of view," he proclaimed, "is the twin brother of Alsace. It has the same climate, the same soil, a similar language, and almost identical customs."[23] Bringing together a cross-border community of scientists, the society's journal requested article submissions from all persons—including French, German, Swiss, and Austrian—researching the nationally ambiguous "Vosges-Rhine" area. Kirschleger's own guidebook for the Vosges, which he wrote for fellow "botanist-geographers," cited studies published by German scientists in his comparison between Vosges flower varieties and those of the Black Forest in Baden. Promoting a different mentality toward Alsatian geography from the explicitly nationalistic Alsatian historical society, Kirschleger's initiatives showed how the French-German border remained a relatively fluid geographic marker for certain members of Alsatian society well into the nineteenth century.[24]

France's defeat in 1871, however, dealt a severe blow to the Second Empire's flourishing historical and naturalist associations in Alsace. The Vosges-Rhine Society ceased to exist altogether, and the Society for the Conservation of Historical Monuments in Alsace lost the majority of its wealthy French-speaking members to emigration. But a group of German elites soon took the place of the French emigrants: professors, schoolteachers, lawyers, judges, railroad engineers, and army officers. Occupying a variety of official posts, the German immigrants were the hands and feet behind Alsace-Lorraine's administrative and military transition from French to German rule. But in spite of their positions within the imperial state, the German immigrants understood that nation building in Alsace-Lorraine required grassroots organization. Thanks to the private initiatives of the leading

cultural, economic, and political class in Alsace, associational life flourished once again in the Vosges. As one German immigrant remembered,
the newcomers to Alsace were "taken in by the rich history of the land and
its magnificent nature with open eyes and open mind."[25] One association
in particular, the Vogesenclub (Vosges Club), brought together the most
powerful German functionaries in Alsace-Lorraine to pursue the common
goal of transforming the Vosges into a German national landscape.

LANDSCAPE BEAUTIFICATION IN GERMANY'S ALSACE-LORRAINE

Following the annexation of Alsace-Lorraine in 1871, German professionals
and bureaucrats arrived in the borderland with dreams of re-Germanizing
their empire's westernmost territory. In their minds, Alsace was a historically German region that had been wrongfully detached from the Holy
Roman Empire in 1648 and subjected to centuries of domineering French
rule. But even though the Germans had at last succeeded in reattaching
Alsace-Lorraine to their state territory, there remained the problem of at-
taching the region to Germany's imagined *Kulturlandschaft*, or cultural
landscape. Because the Francophile Alsatian historical society had already
surveyed its region's landscape for sites of "French national patrimony,"
the newly arrived German immigrants had to figure out how to remap the
region to reflect a German national gaze. They did so using methods similar to those of the French—selecting and cataloging a visual archive and
promoting their own version of an "outdoor museum." But the Germans
took the establishment of a nationalized Vosges landscape a step further
than the French. They would modify the mountain terrain itself in order
to make their "outdoor museum" accessible to large numbers of hikers and
tourists.

Founded in 1872 by Richard Stieve, a German immigrant working as
a magistrate in Saverne, the Vogesenclub was Alsace-Lorraine's first hiking club (and today the largest civil association in eastern France, with
thirty thousand members). The club's stated goal was to turn the Vosges
wilderness into a safely navigable space by clearing and maintaining a network of hiking trails, hammering in directional signposts, installing public
benches, constructing dams and bridges over streams, and erecting panoramic viewing towers. Like an English gardener cutting an ordered set of
hedges out of a wild entanglement of overgrown plants, the Vogesenclub
groomed and managed the natural environment of the Vosges to provide
tourists with a reliable system of navigation and access to picturesque sight
lines. Club members referred to their activities as *Landschaftliche Verschöne-
rungsarbeitung*, or landscape beautification, a concept popular among hiking clubs throughout the German Empire.[26] Landscape beautification di-

rectly served the club's goal of constructing a culturally German landscape in the Vosges. The access that the club opened up to the mountains' historical and picturesque sites (together with expansions in rail travel) brought more people in contact with the Vosges during the late nineteenth century than ever before, turning landscape tourism into a symbolic rite of passage for entering into the German national community.

The Vogesenclub's ambitious goal of "beautifying" a mountainous terrain covering nearly a third of Alsace-Lorraine's surface area required a large, coordinated group effort. Within just a few years, however, the club was able to gather together enough men (and a few women) to accomplish their goal of rendering the Vosges accessible for mass tourism. Club members, mostly German immigrants, viewed the care and management of Alsace-Lorraine's historic German landscape as an act of civic virtue. Even if Alsace-Lorraine was not their native *Heimat*, their fervent belief that the borderland was an *urdeutsch* region—a territory with ancestral German roots—inspired their search for visual and tangible markers of the region's German national patrimony. Reviving the long-forgotten medieval German name for the mountains—the Wasigen—the club celebrated achievements and opened council meetings with the cheer: "Wasigen hoch!" (Cheers to the Wasigen!). Unlike Germans situated in Berlin, who relied on topographical and statistical maps for their territorial knowledge of Alsace-Lorraine, Vogesenclub members cultivated an intimate, experiential understanding of the borderland's geography. Facilitating hands-on contact with the local environment, the Vogesenclub had a decentralized administrative structure. The majority of the club's construction projects and social activities revolved around dozens of local sections, each of which was responsible for carrying out and financing its own "landscape beautification" activities. By 1914, there were fifty-nine such sections, with a total membership of 8,801 people.[27] The German state had only minor influences on the club's work, and the German Forest Administration gave the club permission to organize trail clearing through public forests. "This is a private affair," insisted the club's German founder. "The state can join as a second line of support. But even without state help, we can achieve many things through unified strength."[28]

Thanks to the Vogesenclub's landscaping efforts, hikers gained easy access to the picturesque and historic sites that club members considered "witnesses to the German history of Alsace."[29] In directing the viewer's gaze toward particular landmarks on the mountain terrain, the club's hiking trails reinforced the notion of a landscape as a "presented space." The first German visual survey of Alsace, a photo album sponsored by the grand duke of Baden in 1874, documented fifty-two such sites. "There is no German-speaking area that is so close and so new to us as the blessed land of Alsace," proclaimed the Vogesenclub president, Julius Euting, in his

Figure 5.3 Image of a ruined castle near Hohbarr from a photography album sponsored by the grand duke of Baden shortly after the German annexation of Alsace-Lorraine. G. M. Eckert, *Bilder aus dem Elsass* (Heidelberg: Verlag von Fr. Bassermann, 1874). Photo. et coll. BNU Strasbourg.

introduction to the photo album. "The goal of this undertaking is to bring [Alsace] closer to our eyes and to make known its particular charms."[30] An emerging visual medium, photography lent a documentary quality to the geographic survey of picturesque wonders in Alsace. The album's photograph of Hohbarr, for example, presented a curious scene in which locals went about their daily lives as the ruins of a medieval castle protruded quite dramatically from the center of the village (fig. 5.3).[31] The historical-literary branch (Historisch-Literarischen Zweigverein), one of the most popular sections of the Vogesenclub, with twenty-five hundred members, played an important role in educating the public about the different castles and ruins pictured in the grand duke of Baden's photo album. Section members organized lectures and published articles on topics such as the origins of the language border between Alsace and France, the early Germanic tribes from the Vosges, and the culture of knighthood in Alsace during the Holy Roman Empire. In 1881, Curt Mündel, a German immigrant from Glogau (now part of Poland), published a guidebook to the Vosges in collaboration with club members Julius Euting and August Schricker that promoted this historical reading of the Vosges landscape.[32]

As the local sections of the Vogesenclub cleared trails, nailed in signposts, and performed other acts of landscape beautification to facilitate public access to the ancient castles and ruins in the Vosges, club leaders began to think about mapmaking. Before the German annexation of Alsace-Lorraine, there were no maps of the Vosges that catered to tourists. Guidebooks had simply reprinted sections from the French Army's *Carte d'État-Major*. Recognizing the failures of state-produced maps to meet the needs of the cultured hiker, club president Julius Euting decided that the Vogesenclub would create its own maps of the mountains. In 1876, Euting himself invented a set of conventional signs for the maps (fig. 5.4).[33] The symbols reflected the accomplishments of the club's landscape beautification work: markings for three different kinds of hiking trails (*einfacher, doppelter, dreifacher*), benches for resting (*Ruhebank*), and shelters for bad weather (*Hütte, Pavillon, Schutzhaus*). The list also included a number of symbols to mark a German cultural landscape: castle (*Schloss*), ruin (*Ru-*

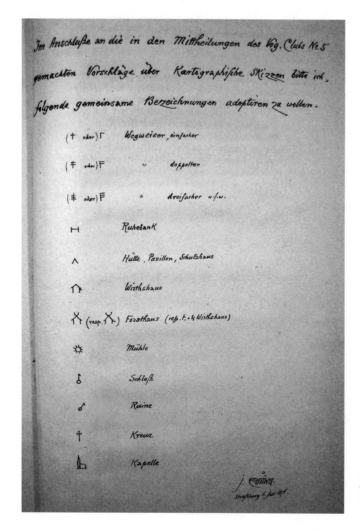

Figure 5.4 Set of symbols for the Vogesenclub's hiking maps. Reproduced from *Mittheilungen aus dem Vogesenclub*, 20 June 1876. Photo. et coll. BNU Strasbourg.

ine), cross (*Kreuz*), and chapel (*Kapelle*). Euting mailed the list of conventional signs to the local sections of the Vogesenclub, asking the members to send him a map of their surrounding area (taken from sections of the *Map of France*) that indicated sites of cultural interest and completed landscape beautification work. When it became clear that local sections had insufficient artistic talent to draw the maps, the club decided instead to publish a professional map series. This map series represented the crowning achievement of the Vogesenclub's years of arduous labor. There was no greater way to reinforce their "mental map" of the Vosges than to create a real map to guide tourists and hikers through their beautified landscape.

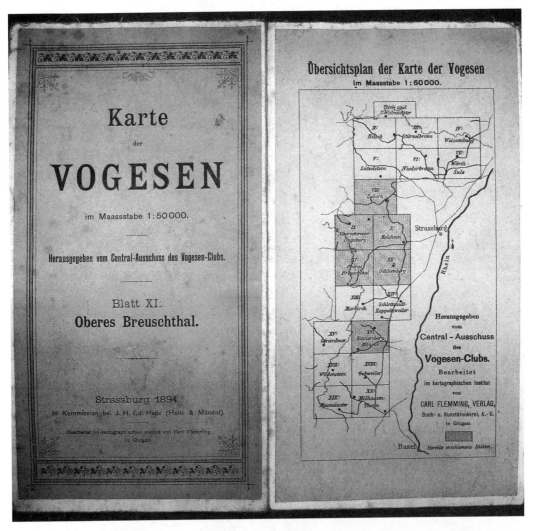

Figure 5.5 Example of a foldable Vogesenclub hiking map with cloth backing. Central Ausschuss des Vogesen-Clubs, *Karte der Vogesen, Blatt XI: Oberes Breuschthal* (Strassburg: J. H. Ed. Heitz, 1894). Fonds patrimonial des médiathèques de Strasbourg.

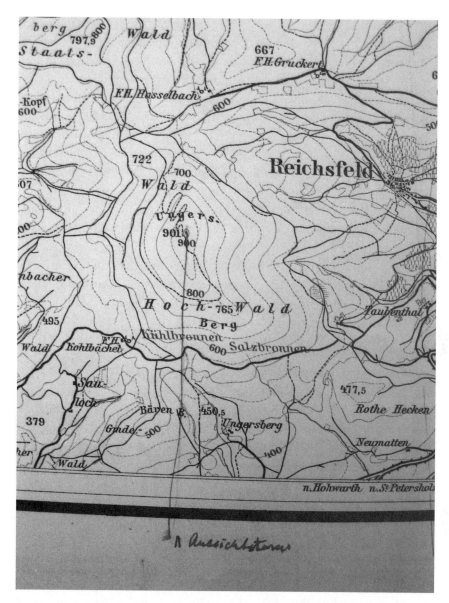

Figure 5.6 Detail from a map "correction sheet" indicates how the Vogesenclub transformed the Vosges Mountain landscape. Here, the map editor's handwritten arrow points to a particular peak and suggests the construction of an *Aussichtsturm*, or viewing tower. *Karte der Vogesen, Blatt XVI: Kaysersberg-Münster. Correcturblatt I.* 1898. Fonds patrimonial des médiathèques de Strasbourg.

From 1894 to 1913, the Vogesenclub published the first "civilian-friendly" map series of the Vosges.[34] For these portable maps, the club chose the scale adopted by the German Tourist Association (1:50,000), large enough to mark hiking trails but small enough to provide a comprehensive overview of the mountain topography. The club created the maps from enlarged *Messtischblätter* (1:25,000) of Alsace-Lorraine, which the Prussian army had surveyed during the 1870s and 1880s. Modifying the state's military and administrative tool for its own purposes, the hiking club added a soft cloth backing for easy outdoor use and replaced military notations with tourist-oriented conventional signs (adopted from President Euting's system). By 1897, the club's maps displayed a vast system of hiking trails, divided into eight *Wegebezirke*, or trail districts, and organized according to one main route (colored red on the maps), secondary routes (colored white and red) and local routes (colored yellow, blue, green and white). A correction sheet (*Correcturblatt*) for the map of the Kaysersberg-Münster area, dating from approximately 1898, underscored how the club's transformation of the Vosges was a constant work in progress (see fig. 5.6). The correction sheet was littered with arrows pointing to the new viewing towers, forest refuges, and hiking paths that the Vogesenclub had constructed, or was planning to construct, in the border area since the map's previous edition. To better understand how the changes indicated on the Kaysersberg-Münster map transformed the outdoor experiences of hikers, we shall examine in detail one of the visual innovations that landscape beautification achieved: the panoramic view.

SYMBOLIC POSSESSION THROUGH PANORAMIC VIEWS

One of the chief goals of the Vogesenclub's landscape beautification was to provide hikers with access to spectacular panoramic views of the Vosges. Panoramic views and viewing towers created expansive sight lines that reinforced the Vogesenclub's understanding of the picturesque—already evident in its trail-clearing and landscape beautification work—as a form of domination and control over the physical environment. According to Stephan Oettermann, the commanding views provided by panoramas allow the viewer to engage in a "visual appropriation of nature."[35] In this regard, the club's panoramas were not unlike the visual perspectives that French and German military surveyors sought when they created their topographical maps. To complete their triangulation measurements, surveyors either turned to existing high points in the terrain, such as church steeples and windmills, or constructed their own wooden signal towers. From this elevated position, surveyors used machines to measure angles and distances in relation to other triangulation points distributed across the terrain. The only difference was that the vogesenclub's panoramas served no measurement function; the view itself was the purpose.

The word panorama, invented in late eighteenth-century England, came from the Greek words *pan* (all) and *horama* (view).[36] It was a term used to describe a form of landscape painting that produced a 360-degree image. During the nineteenth century, city dwellers across Europe paid to visit walk-in panoramas, where they could experience all-encompassing views of landscapes and battle scenes created from the circular arrangement of large-format paintings. For hikers and nature lovers visiting Europe's mountains, printed panoramas were a useful supplement to maps, whose aerial perspective often left the viewer with a feeling of disconnect from the represented environment.[37] Thanks to the panorama's horizontal view, drawn from a single vantage point, the viewer could easily imagine him- or herself within the panorama's field of vision, creating the humanized perspective on landscape that hikers and tourists favored. Because their commanding views of the horizon gave people the feeling that the world was organized around them, historians often associate panoramas with a bourgeois view of the world in which man dominates over nature.[38] In the context of late nineteenth-century European politics, such dominating views of nature not only reflected this bourgeois worldview, but also reinforced the competition for national possession over territory.

During the 1880s and 1890s, the Vogesenclub sponsored the publication of several printed panoramas, each approximately one meter in length, taken from the highest peaks of the Vosges. These printed panoramas provide valuable insight into the national politics framing the tourist experience on Vosges mountaintops during the late nineteenth century.[39] The artist who drew the panoramas was Julius Naeher, an engineer and building inspector from Baden who edited a journal on tourism in southwest Germany. Naeher completed panoramas of several high points in the Vosges, including the Hohneck and Mont St. Odile, but his panorama of the Donon crest offers a particularly interesting connection to previous French landscapes in the region.[40] Like many other historic landmarks in the Vosges, the Donon crest became accessible to tourists only in the 1880s, when the Vogesenclub cleared a trail up to the mountaintop from Schirmeck, a town at the base of the mountain recently serviced with a railway station. Describing the panorama, Naeher called the Donon "one of the most beautiful and historically interesting viewpoints in the North Vosges."[41] In designing his image of the landscape, Naeher followed the standard techniques of a nineteenth-century panorama artist. He drew the image from a single vantage point, according to careful observations from the field, not unlike a topographical map surveyor. To provide the viewer with a sense of distance and scale, he created a foreground rich in details and added images of people. The result was a commanding view of the horizon: a vast open space without fences or borders.

Naeher's panorama illustrated how commanding views of land could acquire different meanings depending on their cultural and political context. By presenting the view from the Donon crest, a site that had previously received much attention from French scholars of Alsace, Naeher symbolically passed the mantle of Vosges historical knowledge from the French to the Germans. Since the seventeenth century, French archaeologists and

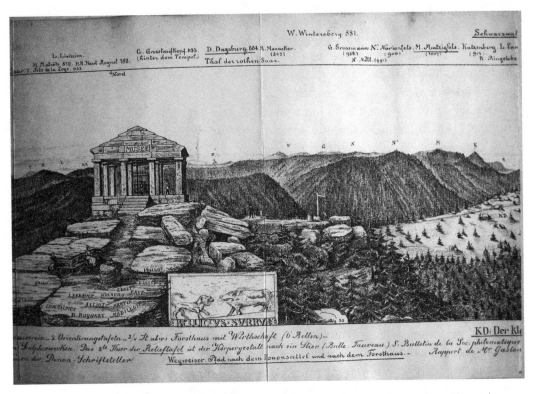

Figure 5.7 In contrast to a map's aerial view, panoramas offered a horizontal view of the landscape, drawn from a single vantage point, so the traveler could easily imagine him- or herself within the panorama's field of vision. Pictured is a section from Naeher's panorama from Donon Peak in the Vosges. The entire panorama measures several feet long. J. Naeher, *Panorama vom Donon im Elsass* (Strassburg im Elsass: Verlag von J. H. Ed. Heitz, 1888). Photo. et coll. BNU Strasbourg.

antiquarians had accumulated evidence of the Donon's past as a "sacred mountain": one of several magnetically vibrating locations in the Vosges where the Gallo-Romans worshipped and made sacrifices to their gods. Because the mountain had served as a ceremonial site, it was an archaeological treasure trove of Roman statues of the god Mercury, as well as fragments of urns, bricks, and coals.[42] Perhaps the most significant French discovery at the Donon, however, was the remains of a rare Gallo-Roman temple. During the 1860s, when Alsatian preservationists and historians were canvassing their region for sites of French national patrimony, they became particularly enamored of its ruins. The Alsatian Archaeology Society, an association researching the region's historic landscape, raised private funds to bring the destroyed Gallo-Roman temple back to life, making the ancient history of Alsace a visible part of its nineteenth-century topography. The building pictured in Naeher's panorama was the reconstructed French temple, completed in 1869, with the word *Musée* on its lintel, reminiscent of the popular "open air museum" concept of its time. Naeher's panoramic view, celebrating the same Gallo-Roman heritage, symbolically appropriated the edifice from the French researchers by adding the names of prominent Vogesenclub members—Schricker, Mündel, and Euting—to the stone walkway in the foreground of the temple.

In addition to the politics surrounding the historic temple pictured in the image, the miniature people that Naeher included in his panorama offer insight into the nineteenth-century culture of nationalized landscape viewership. Unlike the single gentleman that Caspar David Friedrich portrayed in *Wanderer above a Sea of Fog*, the figures in Naeher's panorama demonstrated how landscape tourism had evolved from an experience in self-cultivation to a group activity. On the one hand, the people in Naeher's panorama are detached from one another, each seeming to peer at a different part of the landscape. The woman and the man in the foreground, for example, are looking in opposite directions; he is using one of the two orientation tables that the Vogesenclub built at the site, and she is sitting with her dog on a bench. Another man is perched inside the Roman temple, looking at the historic fragments, perhaps imagining the ancient rituals that had taken place there. In spite of their separate viewing activities, however, the people pictured at the Donon were engaged in a form of group social activity. A train car brought them to the Donon together, they used the same Vogesenclub trail to reach the summit, and they probably read identical descriptions of the Donon site in the *Mündel*, the Vogesenclub's recommended guidebook. During the late nineteenth century, the personal, meditative experience of landscape viewing was thus increasingly structured by collective transportation, literature, and associational culture.

To underscore the importance of cultural context for understanding the meaning of panoramic views, it is helpful to compare Naeher's panorama of the Donon crest to a Frenchman's description of the same panorama from 1894. In contrast to the culture of celebratory national conquest surrounding German panoramic views in Alsace-Lorraine, the Frenchman, Gustave Fraipont, presented the Donon's commanding visual sight lines through a narrative of mourning, loss, and longing:

> The panoramas are completely beautiful: when you go to admire them, you will do as we did: if the weather is clear like today, you will come and seat yourselves on one of the steps of the temple, and fix your eyes toward Alsace, which at this moment sends small billows of smoke, pushed by a light breeze, into France. You will find there, there at the bottom, the elegant spire of the Cathedral of Strasbourg. You will tell yourselves that the needle on its clock moves like the needles on French clocks and that, like the beats of two hearts that love each other, their tick tock beats in unison.[43]

Fraipont's description of the Alsatian landscape that he saw from the Donon crest was clearly embellished. For example, wind patterns in Alsace would have sent smoke billowing in the direction of Germany, toward the east, not westward toward France. Nonetheless, in referring to the sensory perceptions of the landscape viewer—the smell of the smoke, the feel of the breeze, and the sound of the cathedral chime—Fraipont framed his visit to the Donon as an intimate personal experience. Reinforcing the painful memory of 1871, Fraipont set up a contrast between the emotional closeness he felt to Alsace-Lorraine from the steps of the Donon and the reality of Alsace-Lorraine's physical separation from France by a national border.

Over time, the Vogesenclub became even more ambitious in its search for triumphal views over Alsace-Lorraine. For the club's twenty-fifth anniversary in 1899, members decided to build a panoramic viewing and orientation tower at Hohfeld (today the Champ du Feu), the highest point of the Vosges in Lower Alsace.[44] The club named its structure the Hohenlohe Tower in honor of two members of the noble Württemberg family who had served as the unelected governor and prefect of Alsace-Lorraine.[45] A looming symbol of German dominance over the Vosges landscape, the Hohenlohe Tower provided tourists with a 360-degree observation platform, accessible from a spiral staircase with 116 steps. Though the tower provided tourists with a romantic glimpse over the wonders of nature, the tower

Nach einer Original-Aufnahme von G. Schmitt, Postkarten-Verlag in Schirmeck.

Figure 5.8 In commemoration of the twenty-fifth anniversary of the Vogesenclub, members built a panoramic viewing tower at Hohfeld (today the Champ du Feu), the highest point of the Vosges in Lower Alsace. Image from *Mittheilungen aus dem Vogesenclub*, 31 October 1899. Photo. et coll. BNU Strasbourg.

itself stood as a symbol of man's domination over nature in an industrial age. The tower's design exhibited a precocious modernity: ironwork was used not just to support the inner structure of the building's concrete walls, but also as a decorative material on the tower's door and staircase, thereby linking the tower to the modernity of the iron railroad. Furthermore, using scientific knowledge to combat the elements, the Vogesenclub equipped the tower with a copper lightning conductor to protect it against lightning strikes.

The Vogesenclub held its inauguration ceremony for the Hohenlohe Tower on a stormy afternoon in the fall of 1898. Leading the way from the nearest train station to the tower was a military band of hussars from the garrison in Strasbourg. Upon reaching the tower, President Euting thanked the imperial prefect of Alsace-Lorraine, Prince Hohenlohe-Langenburg, for his moral and financial support. In honor of the club's twenty-five years of civic service, the German imperial government had agreed to pay for two-thirds of the tower's costs.[46] The audience, composed of some one hundred members of the Vogesenclub, exclaimed the club salute "Wasigen Hoch!" three times, followed by the German songwriter Christian Schmitt's "Vogesenlied," sung to the melody of the patriotic "Strömt herbei, ihr Völkerscharen." Singing was a frequent activity at club activities, and this moment was no exception. According to the Vogesenclub journal, a number of Alsatian villagers watched the event curiously from the sidelines. Following the ceremony, the Vogesenclub organized a dinner for the celebrants at a local tourist hotel, where Prince Hohenlohe-Langenburg led a toast to the Emperor Wilhelm II.[47] The completion of a new panoramic view was thus a festive occasion to celebrate Germany's symbolic possession of Alsace-Lorraine.

MAPLESS TRAVELERS: AN ALSATIAN REGIONALIST VISION FOR THE VOSGES

The Vogesenclub hoped that its public service work, for which it was commended at the Hohenlohe Tower ceremony, would serve as a much-needed source of reconciliation between the German immigrants to Alsace-Lorraine (the so-called Old Germans) and native Alsatians and Lorrainers (the so-called New Germans). It was true that Alsatians traveled to the Vosges in greater numbers after the German annexation than before, in part thanks to the Vogesenclub's trail-making efforts. Nonetheless, the majority of the Vogesenclub members, and virtually its entire leadership, would remain German immigrants up through 1914. In the final years of the German annexation, however, the club's nationally oriented landscape tourism in the Vosges faced a challenge from another kind of tourism that reflected the rise of regional pride in Alsace-Lorraine. In the face of the German immigrants' push to create a national *Kulturlandschaft* in the Vosges, regionalist-minded Alsatians constructed an alternate vision of the Vosges landscape steeped in the imagery of homeland, or *Heimat*.

Shortly before World War I, landscape tourism in the Vosges reflected the tensions in Alsace-Lorraine's civil society between a dominant class of German-born professionals and functionaries and a group of pro-French, regionalist-minded Alsatian elites. In 1908, over beers at the Taverne Alsacienne in Strasbourg, a group of Alsatian regionalists decided to reclaim their local landscape symbolically from the German nationalists by founding their own hiking club: the Touring-Club Vosgien. Club members stated that their "small but valiant Touring-Club Vosgien" was open to "all true Alsatians [a pointed criticism against the immigrants] who love their region and have just one goal: to get to know it and to love it even more."[48] Calling their journal *Nos Vosges* (Our Vosges), the members of the regionalist club framed their wanderings through the Vosges landscape as an experience in local culture. What made the Vosges home for the regionalists was not the "German" history of the castles, ruins, and medieval villages, but rather the feeling of walking across the land of their forefathers and exploring the small streams and tree stumps which had brought them joy in their youth. The stunning 360-degree panoramic views offered by the Hohenlohe Tower did not interest the Alsatian regionalists. They did not need to achieve symbolic domination of a place that was simply their home.

Members of the Touring-Club Vosgien resented the notion that German immigrants could develop an intimate territorial understanding of their region while the Alsatians themselves lagged behind. Indeed, Vogesenclub members had initially believed the Alsatians to have a "limited knowledge of their own region."[49] The German-led club's landscape beautification efforts were in part intended to educate the native Alsatians about sites of picturesque and historic significance that they had ignored in the past. Defending themselves, the regionalists took aim at the immigrants' methods for landscape beautification, mocking their large-scale administrative organization of terrain into hiking districts and color-coded paths. Calling them *Pfadfinder*, or boyscouts, the club made fun of the German hikers for their artificial relationship with the Vosges Mountains. The Alsatian cartoonist Hansi's satirical images of Vogesenclub members with round red noses in full hiking gear echoed the regionalist club's description of the Germans as people who traveled to the Vosges with "sandwiches, blankets, and tents, with their naive illusion of discovering trails marked and indicated by others."[50] Rather than rely on marked trails, the Alsatian regionalists preferred less structured methods of travel, favoring local guides who knew the forests like the back of their hand. They believed that someone with a real knowledge of the local environment, or *Ortskenntnis*, did not need maps or

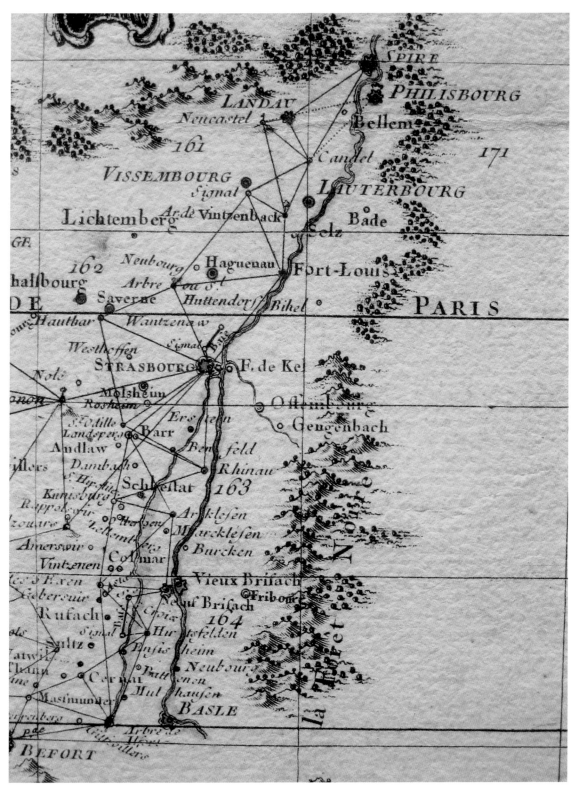

Plate 1 Detail from Cassini III's map of the "Great Triangles." The meridian of Paris stretches across the center of Alsace, demonstrating that even the most remote border areas of France were spatially oriented around the state capital. Bibliothèque nationale de France, Ge BB 565 A (VII), plate 10.

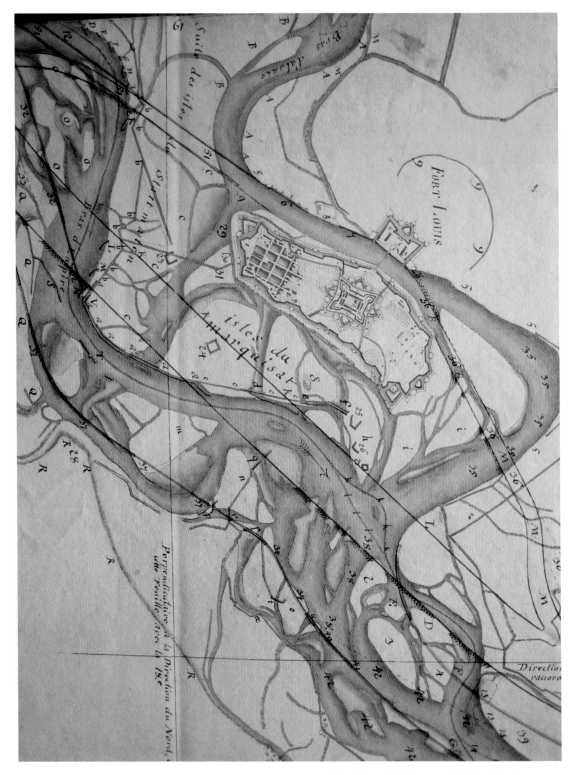

Plate 2 The French state attempted to "fence in" its territory through scientifically measured borders. The military surveyor Jean-Claude Le Michaud d'Arçon soon discovered that this was easier said than done along the Rhine. *Carte des frontières de l'est*. Bibliothèque nationale de France, Ge DD 5464 (17).

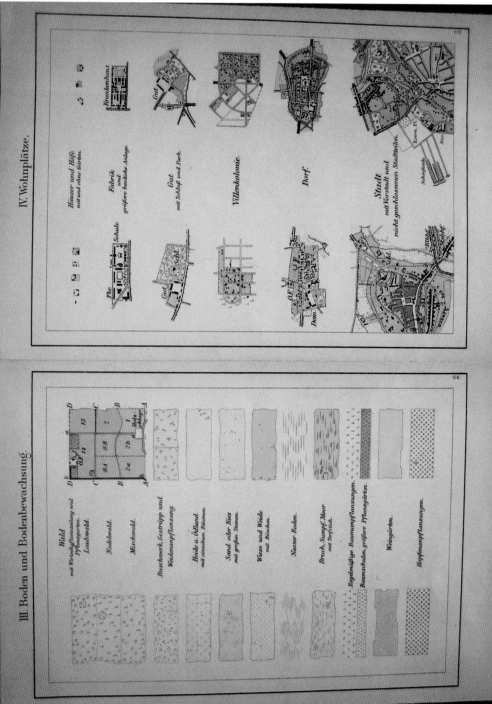

Plate 3 Maps of the German Empire shared a standardized set of symbols and graphics. Sample image from the *Musterblatt und Zeichenerklärung für die topographischen und kartographischen Arbeiten im Masstabe 1:25,000* (Berlin: Königlich Preussische Landesaufname, 1913). Reproduced with permission from the Staatsbibliothek zu Berlin- Preussischer Kulturbesitz.

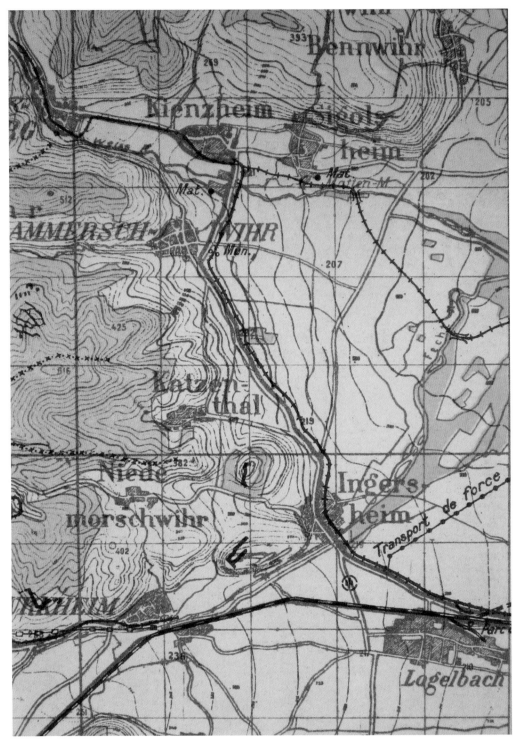

Plate 4 Example of a French *plan directeur,* or battlefield map, from World War I. The French Army converted the map from a German *Messtischblatt* (1:25,000) of Alsace-Lorraine for their military operations in the Vosges Mountains during 1918. *Carte secrète de l'État-Major de la région de Sainte-Marie-aux-Mines,* 1 September 1918. Archives départementales du Bas-Rhin, 7K3.

Plate 5 Ethnic boundaries took the place of existing state boundaries in this atlas image from 1846. *Übersicht von Europa mit ethnograph. Begränzung der einzelnen Staaten, und den Völker-Sitzen in der Mitte des 19ten Jahrhunderts,* 1846, in Heinrich Berghaus, *Physikalischer Atlas* (Gotha: Justus Perthes, 1848), vol. 2. David Rumsey Map Collection.

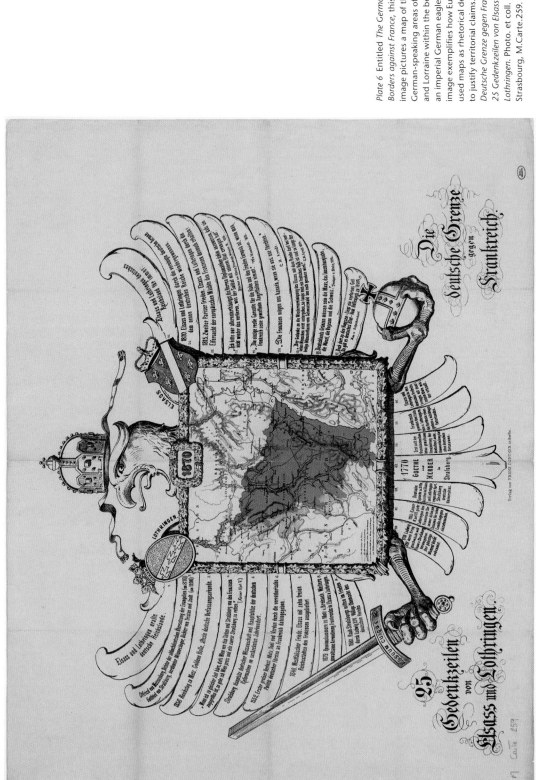

Plate 6 Entitled *The German Borders against France*, this 1872 image pictures a map of the German-speaking areas of Alsace and Lorraine within the belly of an imperial German eagle. The image exemplifies how Europeans used maps as rhetorical devices to justify territorial claims. *Die Deutsche Grenze gegen Frankreich: 25 Gedenkzeilen von Elsass und Lothringen.* Photo. et coll. BNU Strasbourg, M.Carte.259.

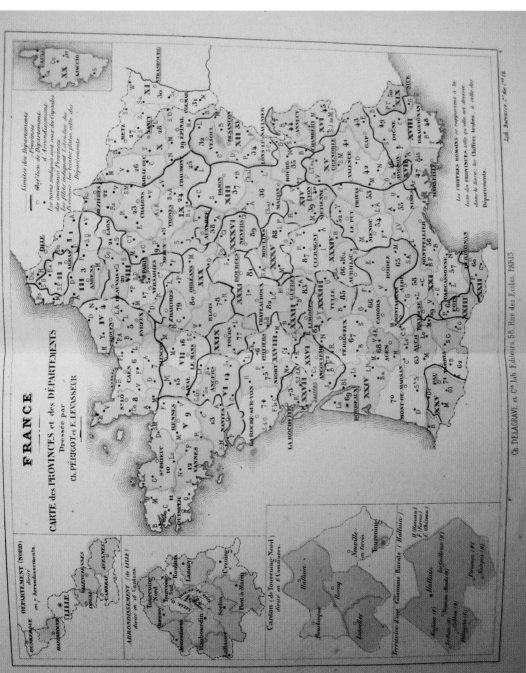

Plate 7 It was a common visual practice among French mapmakers to leave Alsace-Lorraine blank or to color it purple as a symbol of mourning. E. Levasseur and Ch. Périgot, *Cartes pour server à l'intelligence de la France avec ses colonies* (Paris: Ch. Delagrave, 1874). Fonds patrimonial des médiathèques de Strasbourg.

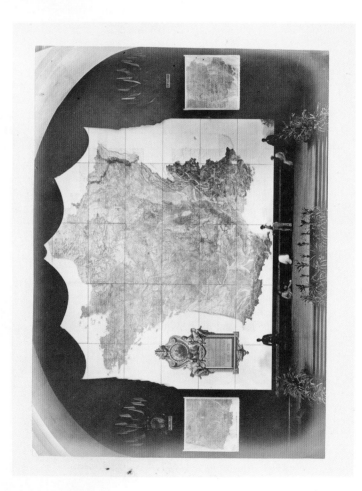

Exposition du congrès international des sciences géographiques Paris 1875
aux Tuileries

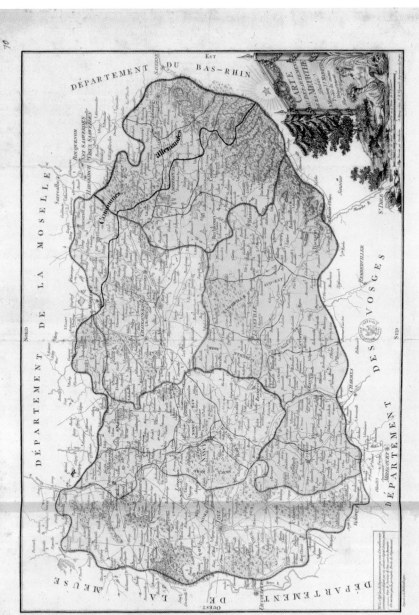

Plate 9 Charles-Étienne Coquebert de Montbret's 1806 map of the French-German language boundary, which traversed the Department of the Meurthe in Lorraine. The path of the language border is indicated with the thick red line on the right-hand side of the image. Note the mapmaker's reference to "German communities" (communes allemandes) on French land. Part of Coquebert's set of maps entitled *Patois de la France: Limites de la langue française*. Bibliothèque nationale de France, NAF 5913.

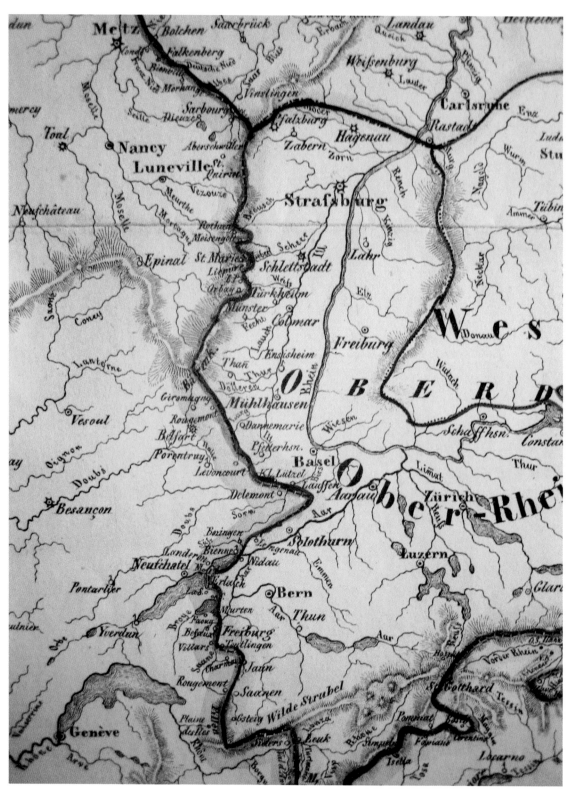

Plate 10 Detail of the French-German language border from Karl Bernhardi's linguistic map of the German Empire from 1844. The map's green language border incorporates Alsace and parts of Lorraine into German-speaking space, while the official border between France and Germany disappears. Karl Bernhardi, *Sprachkarte von Deutschland* (Kassel: Verlag von J. J. Bohné, 1844). Photo. et coll. BNU Strasbourg.

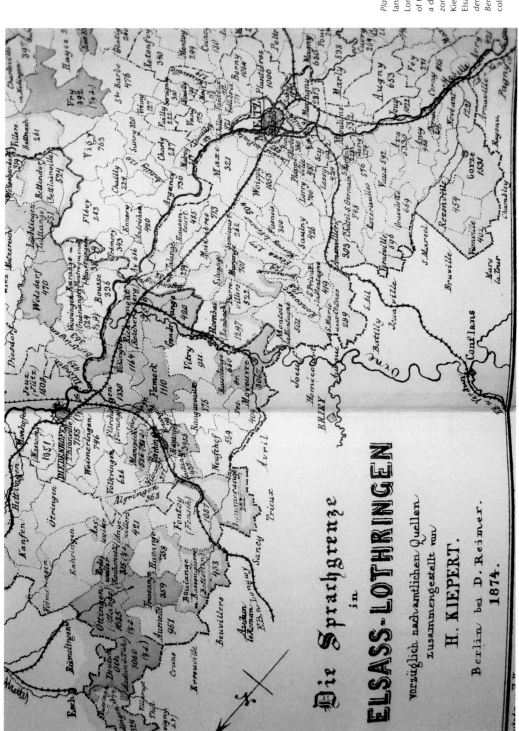

Plate 11 Heinrich Kiepert's language map of Alsace-Lorraine transformed each of the region's districts into a distinct language-speaking zone, or *Sprachgebiet*. Heinrich Kiepert, "Die Sprachgrenze in Elsass-Lothringen," *Zeitschrift der Gesellschaft für Erdkunde zu Berlin* 9 (1874): 316. Photo. et coll. BNU Strasbourg.

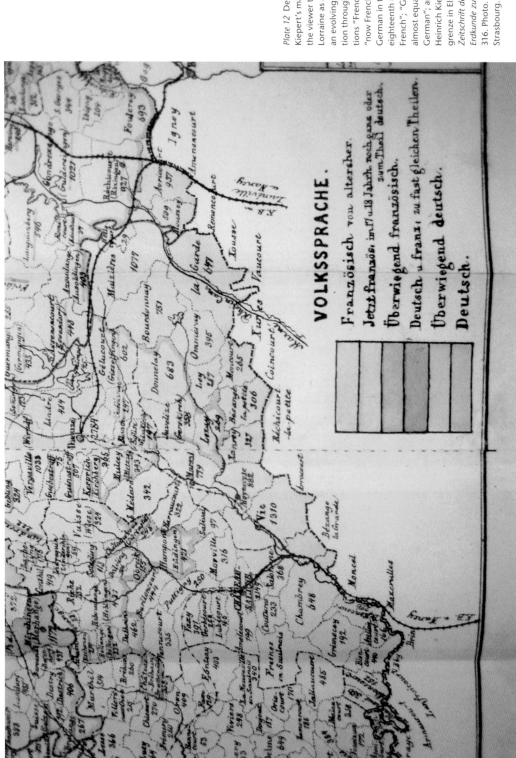

Plate 12 Detail of the key from Kiepert's map that instructs the viewer to visualize Alsace-Lorraine as a borderland with an evolving linguistic composition through the categorizations "French since antiquity"; "now French, partly or entirely German in the seventeenth or eighteenth century"; "mostly French"; "German and French, almost equal parts"; "mostly German"; and "German." Heinrich Kiepert, "Die Sprachgrenze in Elsass-Lothringen," *Zeitschrift der Gesellschaft für Erdkunde zu Berlin* 9 (1874): 316. Photo. et coll. BNU Strasbourg.

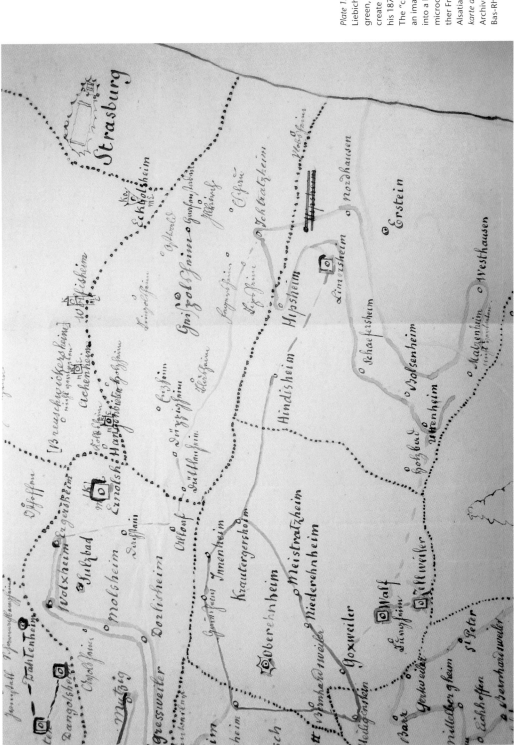

Plate 13 Pastor Louis-Gustave Liebich used yellow, blue, green, and red-colored lines to create language pathways on his 1876 map of middle Alsace. The "counter-map" created an image of land broken up into a kaleidoscopic array of microcultures that were neither French nor German, but Alsatian. *Entwurf einer Sprachkarte des MittelElsasses*, 1876. Archives départementales du Bas-Rhin, 159 AL 546.

Plate 14 This page from the *Atlas national illustré* pictured a map of the Department of the Bas-Rhin surrounded by lithographic designs of local heroes, towns, and bucolic landscapes. Archives départementales du Bas-Rhin, 2K27.

Panorama de Strasbourg. Ville. Page.

PP. 3

Dessiné d'après nature p. F. Piton.

Lith. E. Simon à Strasbourg.

Cathédrale de Strasbourg.

Vue de la Plate-forme et de la Cassine des Gardiens.

37

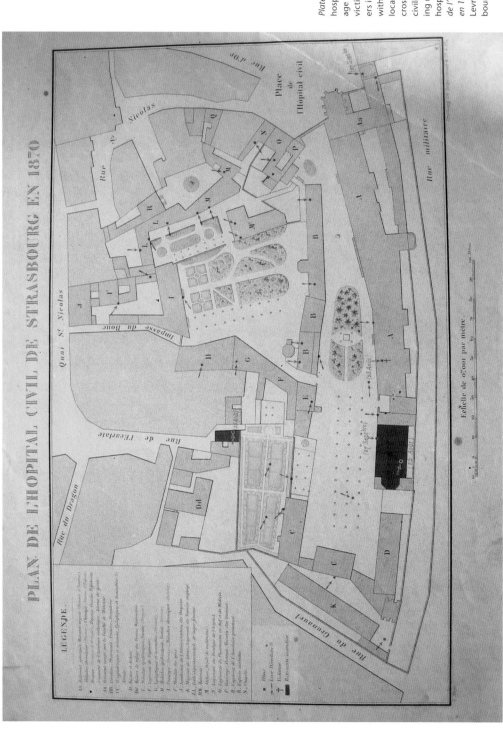

Plate 16 This plan of the city hospital presented an image of Strasbourg as a space victimized by German attackers in 1870. Small red circles with red arrows indicated the location of shell strikes, black crosses marked the site of civilian deaths, and red shading revealed the parts of the hospital that had burned. *Plan de l'hôpital civil de Strasbourg en 1870*, published by Berger-Levrault. Archives de Strasbourg, 272 MW 86.

guidebooks. The description of one hike read: "We prefer to be guided by a member of the club committee, a tireless man who is not scared to climb and does not hesitate at a fork in the road. With him, you only need to walk, without worrying about which direction, your only worry being that you do not keep up with the speed of his steps."[51] Though the Touring-Club Vosgien's membership never exceeded three hundred persons, the association's support for regionalist landscape tourism marked an important step in the development of Alsatian cultural identity.

In terms of visual politics, the Touring-Club Vosgien had arguably created in the Vosges the same *image locale,* or local image, that Alsatian villagers and Strasbourgeois had simultaneously invented in their rural and urban environments. According to historian François Pétry, local images provided a creative space for Alsatians to develop their regional identity in the face of imperial authorities.[52] The *image locale* that the Touring-Club Vosgien envisioned was akin to the image of a big backyard. It was an intimate view of landscape conceived as a family's generational homeland. "What we want is to conserve intact the patrimony that was passed down to us from our fathers," explained the club's secretary, Auguste Muhlberger.[53] After Alsace-Lorraine returned to French rule in 1918, the *image locale*, distinct from the imperial image of a German *Kulturlandschaft*, continued to resonate with many Alsatians. "My soul was imprinted with this first image and has since been molded according to this image," wrote one Alsatian hiker.[54] Though the concept of the *image locale* ran counter to the German nationalists' view of the Vosges as part of a larger German national space, the Touring-Club Vosgien nonetheless owed its entry into landscape tourism to the precedent set by the German hiking club. Under the pressures of the annexation, the Alsatians learned to recognize the value of landscape for creating and sustaining group identity. Following Alsace-Lorraine's transfer to French rule in 1918, however, regionalist landscape imagery would struggle to coexist with the French nationalists' view of the Vosges as a "martyred landscape."

A GERMAN HIKING CLUB BECOMES FRENCH

"The most important organization in Alsace and Lorraine—the Vogesenclub—included, in 1914, nearly 8,000 members, the majority Teutons. Its 56 sections were headed almost exclusively by *boche* functionaries. It is vital to transform these groups without delay into French bases."[55] Following the Allied victory in World War I, the upper echelons of the Vogesenclub fell victim to the French government's purge of German bureaucrats and anyone of German ethnicity. The handful of remaining Alsatian members of the now defunct organization wrote to the French provisional government in Strasbourg, asking for permission to resurrect a French version of the club, arguing that the "boche" organization's landscape beautification work should be continued according to French interests. French authorities agreed, recognizing the club's statutes (an exact replica of the former German club's) on 11 November 1919 and offering its members money to manage trails leading to historical monuments, castles, as well as the World War I cemeteries that now lay scattered across the mountains. Pierre Zuber, a furniture salesman from Strasbourg, became the president of the reborn Club Vosgien, taking over the former German hiking club's 174,000 marks in cash, in addition to its war bonds, property, and copyrights to its maps.[56] Along with several other leading members of the Club Vosgien, Zuber had participated in the regionalist Touring-Club Vosgien during the German annexation. Newly arrived French bureaucrats and administrators from Paris, unlike the Germans before them, were for the most part uninterested in the club's hiking activities.

Despite its new leadership under pro-French Alsatians, the Club Vosgien was essentially a German club under a new name. The Club Vosgien maintained the German club's bureaucratic approach to land management and navigation, appointing an "inspector general" of the club's trail network with six local inspection groups.[57] Through the interventions of Allied authorities, club members succeeded in obtaining the full collection of the German club's lithographic printing stones from Stuttgart, where maps of the Vosges had been printed since 1909. German authorities blocked the map stones' extradition for years, but in 1922, the Club Vosgien was finally granted legal possession of them.[58] The French club published copies of these German-developed maps, with changes only to toponyms and war-damaged sites. The maps would play an important role in the French club's economic success, and President Zuber called the maps "one of the most important and possibly the most useful institutions of the Club Vosgien."[59] In addition to its trail-clearing and mapping work, the Club Vosgien also maintained the former German club's goal of acquiring and disseminating a body of historical and cultural knowledge about the mountains. The Société historique, littéraire et scientifique du Club Vosgien, like its German predecessor, published articles on flora and fauna, local and regional history, and biographies of important Alsatians. In 1922, the hiking club organized a conference on the history of Alsace, where the eminent medievalist Marc Bloch delivered a lecture on the region's history in the ninth and tenth centuries.[60] However, in the wake of the German-led Vogesenclub's intensive turn-of-the-century research into the Germanic roots of the castles and ruins in the Vosges, it was difficult for the French club to completely upend the Germans' claims to the historic monuments. Faced with this situation, the Club Vosgien developed a new definition for the term "picturesque" to go along with the "local image" concept, creating an innovative French vision of the Vosges landscape shaped by the psychological and political needs of the postwar period.

MARTYRED LANDSCAPE

While the Francophile Club Vosgien maintained many of the goals of its German predecessor, it also undertook, in cooperation with veterans' and local village associations, a new type of landscape beautification with an explicitly French national purpose: clearing trails to the World War I battlefields in the Vosges.[61] Unlike the landscape beautification of the turn of the century, which focused on scenic hikes and commanding views, this beautification work challenged the very notion of "beautification," because its aim was to preserve the horrific destruction of the mountains' natural beauty. Recognizing that battlefront visits were drawing thousands of French visitors to the Vosges, the Club Vosgien—many of whose members had fought on the German side during the war—cooperated with associations from interior France to reframe the mountains as a "martyred landscape." Their principal site of interest was Hartmannswillerkopf (also called Vieil Armand), a strategic observation point on a high promontory overlooking the Alsatian Plain. During the war, French and German troops had fought desperately for possession of the southern Vosges mountaintop. The number of Frenchmen who died per meter of front at Hartmannswillerkopf was approximately equivalent to those who died at Verdun and Artois: more than four men per meter of front. In total, sixty thousand French and German soldiers died trying to possess the rocky crest.[62] The remains of many soldiers were never identified, and it became one of the war's largest ossuary sites.

The French government classified the battlefield at Hartmannswillerkopf as a historical monument in February 1921. Shortly thereafter, associations of veterans and local Alsatian villages came together to form the Committee for the Monument at Hartmannswillerkopf, headed by General Georges Tabouis, the commander who had led the French assault on the mountain. Instead of restoring the mountain's natural appearance, the committee decided to preserve Hartmannswillerkopf in its state of destruction in order to symbolize the loss of innocence and the great sacrifice that France had endured to recover Alsace and Lorraine. The committee nominated Robert Danis, the Parisian-appointed director of the Commission for Architecture and Fine Arts in Alsace-Lorraine, to design a memorial at the site. Danis used the destroyed mountain itself as his inspiration. He added only a few small, subtle edifices and a crypt, drawing the viewer's gaze to the mountain's martyred landscape: a surreal land denuded of trees, carved

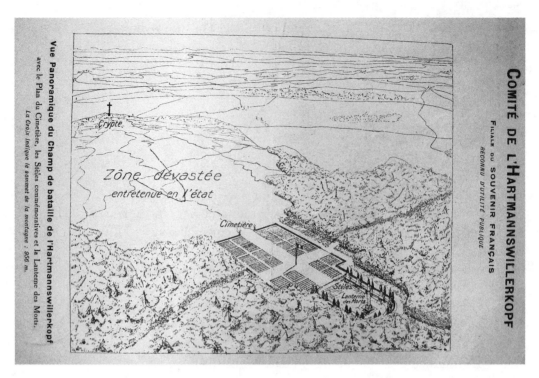

Figure 5.9 In 1921, the French government declared the World War I battlefield at Hartmannswillerkopf, Alsace, a historical monument. Comité de l'Hartmannswillerkopf, "Vue panoramique du champ de bataille de l'Hartmannswillerkopf," 26 March 1922. Archives départementales du Bas-Rhin, 121 AL 1091.

up by trenches, pockmarked by shell explosions, and covered with tangled coils of barbed wire.[63] Commending Danis's artistic vision for the memorial at Hartmannswillerkopf—one of the four major war memorials constructed in France after the Armistice—Marshal Pétain wrote: "It was about making the terrain itself the principal element of the envisioned plans. Mr. Danis, inspired by the great architect Vauban, has proposed nothing more than to proceed with surveys for the regularization of the ground, and to adapt some very simple architectural elements, similar to elements of fortification. In sum, he has wisely adjusted his vision to the configuration of the terrain."[64] Introducing an original form of memorial design for the twentieth century, Danis had creatively used land formations to convey the tragedy of war in the disfigurement of an immaculate mountain.[65]

When Danis's plan for preserving the destroyed terrain of Harmannswillerkopf was complete, the memorial committee created a walking path, visible in the dark line on the battlefield plan, for visitors to make a pilgrimage to the top of the mountain, where they would reach a large cross after passing through the traumatized landscape (fig. 5.10).[66] The committee decided to mark the summit with a cross, a symbol of redemption reigning over the mangled slopes below, because they considered it to be a "sober, beautiful, and pure symbol of sacrifice."[67] They wanted the cross to be large enough to clearly distinguish Hartmannswillerkopf from neighboring summits, so that anyone passing through the Alsatian Plain would look up and exclaim "Voilà l'Hartmann!"[68] Although the Club Vosgien had favored the construction of "pilgrimage" trails leading up to the denuded summit of Hartmannswillerkopf, the club protested against the addition of the giant cross. "As much as our compatriots enthusiastically support the goals of the monument," read a Club Vosgien editorial describing the popular Alsatian reaction to the proposed memorial, "the planned method for creating the monument incites their legitimate objections."[69] On behalf of their membership, the Central Committee of the Club Vosgien wrote to the architect, Robert Danis, complaining against what they described as "a gigantic and colossal cross" that would, to their distress, be "silhouetted on the harmonious crest of our Vosges and disfigure their appearance."[70] In spite of the club's support for the French repossession of Alsace-Lorraine after World War I, it was clear that the Alsatian-led hiking club was wary of establishing too bold a symbol of French national identity on their mountain

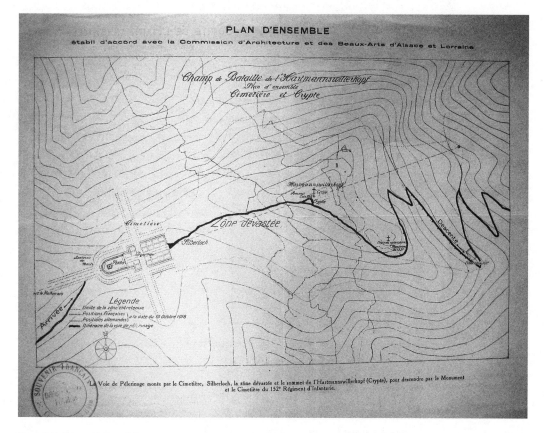

Figure 5.10 This aerial plan of the monument at Hartmannswillerkopf indicates, with a curving dark line, the pilgrimage itinerary that visitors took past the cemetery, through the destroyed battle zone, up to the cross at the summit, and then downward. Comité de l'Hartmannswillerkopf et Comité de la Reconnaissance Alsacienne-Lorraine, "Monument de l'Hartmannswillerkopf." Archives départementales du Bas-Rhin, 121 AL 1091.

landscape. The large cross would have changed the familiar "local image" of their rounded Vosges crests too dramatically. Upon reading the Club Vosgien's protests, Danis reconsidered his plans and agreed to lower the height of the cross from seventy-five to twenty-five meters.

Citizen mapmakers played a leading role in creating innovative national and regional landscapes in the disputed Vosges Mountains. French-, German-, and Alsatian-led civil associations each surveyed the Vosges for sites of historic and picturesque interest, building their own visual archives of the territory's cultural landmarks in the form of maps, lithographic imagery, and photographs. The associations developed successful strategies for sharing their particular geographic vision of the Vosges with the wider public, engaging in the debate over Alsatian identity through visual forms of communication. Whether it was a vision of a French outdoor museum, a German cultural landscape, an Alsatian homeland, or a martyred landscape, associations in Alsace-Lorraine's fractured civil society all used visual images in their struggle to possess the French-German borderland. If we consider the continuities between the visual techniques that French- and German-led associations employed, we can also view the clubs as meaningful points of transnational cultural exchange.

While Alsace's largest urban area would appear to have little in common with its rugged mountainous terrain, patriotic mapmakers employed similar methods for creating national cityscapes in Strasbourg as they used to invent national landscapes in the Vosges. Located directly on the Rhine border between France and Germany, Strasbourg became a site of both conflict and mediation between French and German populations during the nineteenth and twentieth centuries. A small city of just seventy thousand people in 1870, Strasbourg grew to become a major urban capital of nearly one hundred eighty thousand residents in 1910.[71] As Strasbourg expanded physically and took on a new role as the administrative and cultural center of Alsace-Lorraine, its urban spaces became persuasive symbols of French, German, and regional identity. The following chapter turns to Strasbourg's changing urban plans and architecture in order to analyze how cityscapes shaped the politics of identity in a disputed European borderland.

CHAPTER 6

Visualizing Strasbourg

High atop the viewing platform of the Cathedral of Strasbourg, which for centuries stood as the tallest building in Europe, the names of hundreds of eighteenth- and nineteenth-century travelers are etched in stone. The hand-carved signatures, a sort of old-fashioned graffiti, serve as vivid reminders of Strasbourg's long-standing popularity as a stopping place for European sightseers. To reach the top of the cathedral, visitors climbed hundreds of well-worn steps along a tight spiral staircase, an ascent that was not for the faint of heart. But once they reached the top, travelers were rewarded with a sweeping, panoramic view that completely transformed their visual perspective on the urban space and its surroundings. With the help of telescopes, the cathedral's visitors could turn to one side and see far across the Rhine into Germany's Black Forest, and then look toward the other direction and cast their eyes along the silhouetted slopes of the Vosges Mountains. Directly underneath the cathedral, the city of Strasbourg unfolded as a tangle of medieval half-timbered houses, canals, and, by the late nineteenth century, modern tree-lined districts to house the city's burgeoning middle-class population. To a certain extent, the visual conquest of land achieved by travelers gazing out from the Cathedral of Strasbourg was not unlike that of the uniformed military surveyors who clambered to the tops of their trigonometric viewing platforms to make scientific measurements of the French-German border territory. But rather than use their commanding views to produce topographical survey maps, Strasbourg's modern tourists played a key role in recasting Alsace's cultural geography.

The city of Strasbourg, whose name derives from the network of Roman roads that once met there, was a critical space of encounter between modern France and Germany. At times, Strasbourg was the staging ground for tremendously violent encounters between citizens from the two countries, perhaps most memorably during the fifty-day German siege of the city in 1870. But there were other historical moments when Strasbourg served as a crucial point of cross-border mediation and exchange between the French and the Germans, particularly in the area of international commerce. Nineteenth- and twentieth-century Strasbourg was thus a city with many faces. Whether it was under French or German control, the city filled multiple roles as a military fortress, an administrative capital, a trading hub, and a symbol of national and regional cultures. Strasbourg therefore holds within its urban anatomy—its walls, monuments, bridges, and river ports—a vast archive of historical evidence to explain the shifting culture of the French-German border encounter. We can interpret the changing appearance of modern Strasbourg, both in paper images and in its architectural constructions, as a microcosm for the broader conflicts and interconnections among French, Germans, and Alsatians.[1]

The term "cityscape" (*image de la ville*, or *Stadtbild*) can be defined as an artistic representation or pictorial view of a city. Before the rapid spread of nationalism in the nineteenth century, European cultural and political leaders did not concern themselves with defining Strasbourg's cityscape as French, German, or Alsatian. During the era of competitive nationalism, however, manipulating the appearance of Strasbourg's cityscape became increasingly important for patriots seeking to legitimize their claims to the French-German borderland. Analyzing the visual techniques that different interest groups used to frame Strasbourg as a national (or transnational) urban space, this chapter will explore how modern European citizens developed their identities spatially, through their built environments. This chapter will also consider how the evolution of urban space created new social geographies that fostered complex power dynamics among the French, the Germans, and the local population.[2]

A rich variety of visual sources can help us to analyze the relationship between identity formation in disputed Alsace-Lorraine and Strasbourg's spatial transformation.[3] Two categories of visual evidence are particularly useful: two-dimensional images of Strasbourg in the form of maps, panoramas, lithographs, photographs, and drawings, and the three-dimensional built landscape of Strasbourg in the form of architecture. Especially important to understanding the role of Strasbourg within broader French and German visions for Alsace-Lorraine are the urban plans—blueprints for the development of the city—which provide material evidence of nationalists' utopian schemes for the future layout of the border space. In addition to examining urban plans with nationalist motivations, we will also explore how Alsatian regionalists turned to the concept of the "local image" (*image locale*, or *Ortsbild*) to emphasize the distinctiveness of the city's regional identity. According to numerous urban theorists, the construction of urban space offers insight into the collective mentalities of a people and the conflict over social power and influence.[4] As visual narratives of nationalist, regionalist, or internationalist agendas, Strasbourg's competing cityscapes help us to trace shifts in collective mentalities toward the French-German border over time.

EXPERIENTIAL AND AERIAL VIEWS IN STRIEDBECK'S
EIGHTEENTH-CENTURY MAP OF STRASBOURG

During the eighteenth century, the French public was already experimenting with innovative approaches to visualizing urban space. Johannes Striedbeck's 1760 image of Argentina, which took its title from Strasbourg's ancient name, provides a helpful piece of evidence for analyzing the relationship between two popular approaches to looking at cities, one experiential and the other a constructed view from above.[5] In the middle part of his multifaceted image, Striedbeck represented Strasbourg from a horizontal perspective. This profile, or "portrait," view was typical of an older style of urban representation that had been popular in Europe for several centuries. It was designed to communicate what a city looked like from horseback or on foot.[6] Striedbeck emphasized the experiential nature of this visual perspective by including tiny figures, pairs of gentlemen with hats and women with parasols, in the process of taking a stroll around the outskirts of the city. In addition to representing the city on an accessible human scale, the profile view enabled someone looking at the image to identify specific buildings and monuments. For example, Striedbeck gave each of Strasbourg's signature landmarks—the cathedral, smaller churches, the town hall, the hospital, and the gates to the city—a number that corresponded to a label with its name at the bottom of the page.

In contrast to the profile view of the city represented in the center of the image, the aerial-view maps pictured in the image's bottom third made the city appear abstract and dehumanized. Through the scientific mapping technologies of the day, Strasbourg was transported from the real world of pedestrians, trees, fields, walls, and buildings to a flat, two-dimensional plane. There were no humans pictured within the maps that Striedbeck drew, and the city took on a snail-like form with rings of river channels and walls. The ideal position of the modern urban observer, he appeared to argue, was that of the surveyor in the middle of the right-hand side of the image. Pictured with his drawing table, the map surveyor demonstrated the relationship between the view that he was experiencing, which was the profile, or horizontal, image of the city, and the top-down view of the city that he was constructing on paper.[7] Each of the maps pictured at

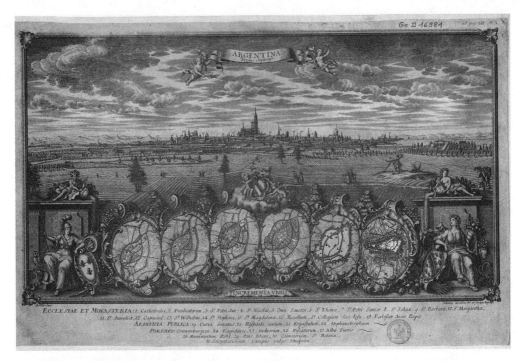

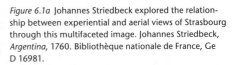

Figure 6.1a Johannes Striedbeck explored the relationship between experiential and aerial views of Strasbourg through this multifaceted image. Johannes Striedbeck, *Argentina*, 1760. Bibliothèque nationale de France, Ge D 16981.

Figure 6.1b In this detail, a surveyor observes the city and uses a drawing table to create his map of Strasbourg. Johannes Striedbeck, *Argentina*, 1760. Bibliothèque nationale de France, Ge D 16981.

Figure 6.1c Striedbeck numbered individual buildings from Strasbourg's cityscape. Johannes Striedbeck, *Argentina*, 1760. Bibliothèque nationale de France, Ge D 16981.

the bottom of the image represented a particular moment in time during Strasbourg's historical evolution from a tiny Roman outpost to a heavily walled modern fortress.

Striedbeck's complex image of Strasbourg signaled the beginnings of a new era in the history of European urban planning, when scientifically surveyed, totalizing views of cities would play an increasingly vital role in shaping utopian ideas about urban space. Rather than maps mirroring reality, reality began to conform to the idealized cities that Europeans represented on maps. But in France, it was difficult to achieve a public consensus on what an ideal French city, and an ideal border city in particular, should look like. Should Strasbourg continue its eighteenth-century role as a heavily walled military fortress, or should it open up to cross-border commerce and exchange with the German lands? Looking at urban plans and images from nineteenth-century Strasbourg, we see that both alternatives were possible.

BETWEEN AN OPEN AND CLOSED CITYSCAPE: STRASBOURG UNDER NINETEENTH-CENTURY FRENCH RULE

During the nineteenth century, representations of France's provincial cities could be found at the center of public view, in one of Paris's most famous urban squares. In 1834, eight massive female statues, each commemorating one of France's great provincial capitals, were built along the octagonal shape of the Place de la Concorde. Though all were sculpted in classical style, two of the statues stood out from the others. Seated on cannons and wielding swords in their hands, the figures representing Strasbourg and Lille embodied the French capital's militarized view of its eastern border. Both cities, heavily fortified since the time of Louis XIV, had played an important role in defending French lands from foreign invasion during the revolutionary and Napoleonic Wars. Even decades later, under the July Monarchy, the memory of these dramatic border conflicts with Central European empires perpetuated a popular French image of Strasbourg as a "fortress city," built to defend the nation against foreign enemies.

In addition to statuary representations of Strasbourg's military importance, urban plans also rendered visible the city's role as a defensive border outpost. While numerous cities in France had shed their exterior walls and ramparts during the first half of the nineteenth century in the interest of economic development, garrison towns such as Strasbourg maintained their imposing fortifications.[8] In going against the broader nineteenth-century trend toward the opening up of city walls, the urban topography of garrison towns such as Strasbourg reflected the French state's continuing mentality toward borders as hostile places that could evolve at any time

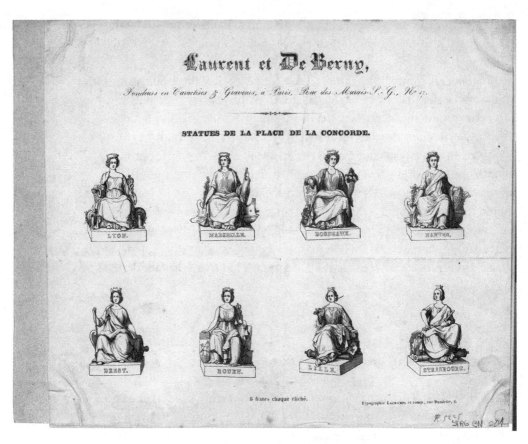

Figure 6.2 In 1834, the city of Paris erected statues representing each of France's major provincial cities at the Place de la Concorde. In this print, the female figure of Strasbourg is pictured in the lower right-hand corner. Laurent et De Berny, *Statues de la Place de la Concorde* (Paris: Lacrampe et co., ca. 1834). Photo. et coll. BNU Strasbourg, STRG.BN.1.

into dangerous war fronts.[9] Strasbourg's tightly bound, closed cityscape pictured in figure 6.3 became synonymous with a tense border relationship with the Germans.

In the realm of commerce, however, there was an evolution during the nineteenth century toward an alternative view of Strasbourg as a point of peaceful contact between France and Central Europe. A flourishing bourgeoisie, committed to a liberal ideology of free trade and industrial economic expansion, sought to move past the old views of frontiers as military barriers and instead channel a border atmosphere of openness and transnational exchange between European lands. This mentality toward borders found its visual expression in images of the French-German border captured from the top of the Cathedral of Strasbourg. The tallest building in the world until 1874, and today the fourth tallest church in the world, Our Lady of Strasbourg offered a territorial perspective on the border cityscape that was far different from the view from below, at the level of the city's fortresslike walls. Soaring above the city, the cathedral's viewing platform presented visitors with a bird's-eye view of the French-German border as an open canvas of land with no clear national divides. Describing what he saw from the cathedral in a letter published in 1842, Victor Hugo wrote:

> In making a tour of the belfry, one sees three mountain chains: the edge of the Black Forest in the North, the Vosges to the West and the Alps in the middle. One is so high that the landscape is no longer a landscape; it is like what I saw from the Heidelberg Mountain, a geographic map, but a living geographic map, with mists, smoke, shadows, and gleam, the quivering of water and leaves, clouds, rain, and rays of sunlight. . . . I would go from one turret to another, admiring one after another France, Switzerland, and Germany in a single ray of sunshine.[10]

Visualizing the borderland below him under a "single ray of sunshine" and portraying foreign landscapes as warm and familiar, Hugo demonstrated to his readers that France's neighbors should be admired and not feared. Hugo was not alone. The view's popularity from the late eighteenth century onward among a wide range of Europeans was evident in the cathedral's ex-

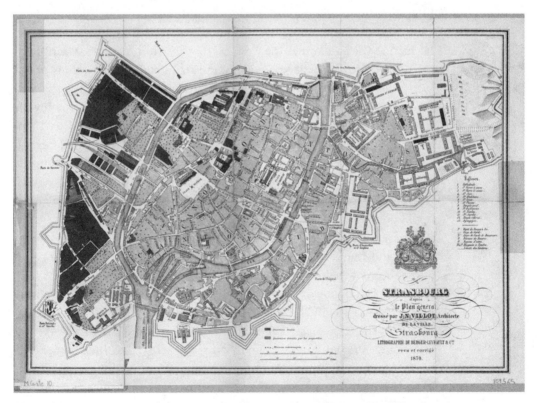

Figure 6.3 Unlike many of France's other provincial capitals, Strasbourg maintained a defensive ring of walls throughout the nineteenth century. Strasbourg's city architect, Jean-Nicolas Villot, drew up this plan of Strasbourg on the eve of the Franco-Prussian War. Jean-Nicolas Villot, *Plan de la ville de Strasbourg* (Strasbourg: Berger-Levrault, 1870). Photo. et coll. BNU Strasbourg, NIM.31985.

tensive "guestbook": hundreds of European visitors (including French, German, British, Polish, Russian, and Italian visitors) hand etched their names with great care into the soft reddish-orange sandstone of the bell tower's walls, where they are still visible today.

The rapid growth of European economies during the nineteenth century strengthened Strasbourg's centuries-old role as a conduit for intellectual and commercial exchange between the right and left banks of the Rhine. As new railway networks snaked their way across Alsatian farmlands and shipping vessels chugged along the recently canalized Rhine River, Strasbourg's bourgeoisie increased its business and social contacts with members of a similar class in the nearby German states. The flourishing middle class's liberal mentality toward the French-German border was documented in a popular guidebook to the city and its surroundings, written and illustrated by the Alsatian Frédéric Piton in 1855.[11] Like Hugo's writings, Piton's panoramic images of Strasbourg, which he drew from the top of the cathedral, presented a border landscape that lacked strict national definition. His engraving of the cathedral's viewing platform pictured men and women in bourgeois dress using telescopes to gaze out toward the French and German lands to the right and left (see plate 15). Framing Strasbourg as the peaceful geographic meeting point of France and Germany, Piton described the view as one of "two mountain chains whose undulating crests cross the horizon to the east and to the west . . . [both] rich in nature's beauties, ruined castles, towns, villages."[12] Echoing Hugo's earlier descriptions of the picturesque cross-border view from the cathedral, Piton's open panorama contrasted dramatically with the ground-level view of the city as a hostile fortress. For example, the curved panoramic sight lines that Piton highlighted with dotted lines on his map of Strasbourg extended well beyond the city's ring of defensive walls, suggesting an openness to the outside world (see fig. 6.4). Thanks to the promotional work of Piton and others, the cathedral platform received high marks from a number of French guidebooks published at the time.[13]

The growing strength of liberal attitudes toward the French-German border and the decreasing importance of Strasbourg's military role could also be seen in the construction of the Rhine Railway Bridge between 1857 and 1861. Just five years after France celebrated the completion of a railway link between Paris and Strasbourg, Emperor Napoleon III collaborated with leaders in the German state of Baden to build a bridge that linked together Strasbourg and Kehl, the German city directly across the Rhine border. The magnificent bridge connecting the cities altered Strasbourg's closed appearance, helping to cement the border encounter between France and the German lands as one of transnational exchange rather than conflict. Its elegant Gothic design symbolized the promise of a new era in which European countries would come together in a vast commercial railway network, in-

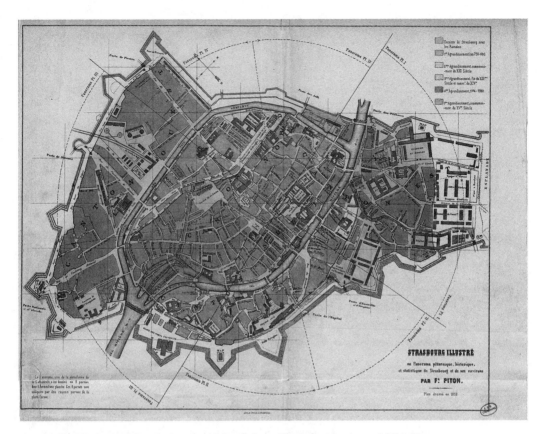

Figure 6.4 The 360-degree perspective from the viewing platform of Our Lady of Strasbourg. Frédéric Piton, *Strasbourg illustré ou panorama pittoresque, historique et statistique de Strasbourg et de ses environs,* vol. 1, *Promenades dans la ville* (Strasbourg: Chez l'auteur, rue du Temple Neuf, 15, et chez les principaux librairies, 1855). Photo. et coll. BNU Strasbourg.

terweaving their economies in a liberal vision of collective prosperity. The bridge's iron structure was not purely functional; its metal towers, fashioned to appear light and airy, reflected a new aesthetic that embraced an emerging European industrial age (fig. 6.5). The mid-nineteenth-century evolution toward an increasingly interconnected French-German border relationship was abruptly shaken off course, however, by the outbreak of the Franco-Prussian War in the summer of 1870. The German armies' fifty-day siege of Strasbourg marked a turning point in modern border politics between France and Germany, bringing an end to the trend toward fluid borders and the beginning a period of violent national conflict.[14]

LANDSCAPE OF DESTRUCTION: THE SIEGE OF STRASBOURG, 1870

How does war change how we see borders? How can visual images structure negative views of enemy territory, dividing interconnected lands into op-

RHEINBRÜCKE.

Lichtdruck von J. Kramer, Kehl.   24   Verlag R. Schultz & Co., Strassburg.

Figure 6.5 From 1857 to 1861, France and Baden worked together to build a modern bridge connecting Strasbourg and Kehl, the German city directly across the Rhine border, bringing their economies closer together. The Gothic-style bridge would become a symbol of the state of relations between French and Germans. Retreating French troops eventually dynamited the bridge in 1940. Entitled *Rheinbrücke*, this image was part of a series of photographs published in J. Kraemer, *Album von Strassburg* (Strasbourg: R. Schultz, 1878). Photo. et coll. BNU Strasbourg, M.116.283.

posing national spaces? Strasbourg's transformation from a site of French-German connection to a site of cross-border conflict from August to September 1870 was immediately apparent in the new visual archetypes of the Rhine border. Drawn from a linear perspective, from the vantage point of the attacker, a German lithograph of the siege of Strasbourg offers an example of an image that framed the Rhine as a gulf separating hostile enemies (fig. 6.6). In the foreground, a German regiment from the town of Kehl lobs cannonballs into Strasbourg, pictured with its medieval Gothic cathedral surrounded by fires emitting black puffs of smoke skyward. On the right-hand side of the image, the damaged Rhine Railway Bridge, which had once symbolized the promise of an interconnected Europe, illustrates the broken state of French-German relations. Popular French images from the Franco-Prussian War echoed this theme of division, likewise framing the Rhine as a rift between hostile neighbors. Gustave Doré's 1870 etching *The German Rhine,* for example, portrayed the German Army as a hoard of savage barbarians emerging from the river with bayonets in their hands.[15] Images of the Rhine as a civilizational dividing line between France and Germany would thus constitute an important visual narrative for the heightened state of border tensions. The transnational landscape images that circulated in the mid-nineteenth century—Victor Hugo and Frédéric Piton's portraits of the French-German border as a continuous geographic canvas—were supplanted by visions of a divided border landscape emphasizing dark views of enemy territory.

Views from the interior of the besieged city likewise signaled the end of Strasbourg's role as a mediating border space. The city bore severe damage from the German shelling; entire neighborhoods became fields of destruction. In spite of the efforts of a volunteer fire corps, many of the city's medieval rooftops went up in flames, sending homeless residents to the quays along the canals in search of temporary shelter. A notice in the local newspaper instructed residents to bury unexploded shells in their backyards.[16] Gustave Fischbach, the journalist who reported on the siege for the *Courrier du Bas-Rhin*, described the pitiful scenes: "The most beautiful streets, the most populated neighborhoods, the outskirts, the public buildings: ruins. The artistic treasures, the scientific collections, the masterpieces and the marvels: dust. On vast expanses, piles of rubble, piles of stones, blackened roofs, twisted metal, debris, fragments: a terrible disorder."[17] Because Strasbourg's gates had to remain closed, the mayor ordered the dead to be bur-

Figure 6.6 This German lithographic image pictures the Rhine as a gulf separating hostile enemies. Note the damaged Rhine Railway Bridge on the right-hand side. *Beschiessung von Strassburg und Kehl.* 1241 N, bpk, Berlin/ Bildarchiv Preussischer Kulturbesitz/ Art Resource, NY.

ied among the exotic plants at the city's botanical garden. Setting the stage for the total wars of the twentieth century, the German military targeted civilian centers, including the city's public hospital, where several deaths were reported. Berger-Levrault, a prominent Strasbourg publishing company, developed an innovative map to publicize the tragic deaths of innocent civilians at the hospital (see plate 16). The map included an overview of the hospital's buildings along with symbols that indicated the direction and location of German shell strikes and black crosses to indicate deaths from shelling. Parts of the hospital that had burned were colored red. In highlighting the ruthless attack on civilians at Strasbourg's hospital, the map portrayed the Germans as enemies of the most coldhearted and brutal nature.

In addition to its civilian centers, Strasbourg's historical patrimony, the joint heritage of French and German cultures, suffered heavy damage from the siege. Shells severely damaged the roof and tower of the cathedral; one witness to the siege described how time itself seemed to stop when the cathedral bell ceased its hourly chimes, leaving the city in an unfamiliar silence.[18] In the most devastating blow to Strasbourg's cultural holdings, on 24 August, German shells struck and destroyed the Protestant Temple-Neuf, which had held the second largest library in France, with four hundred thousand volumes. In addition to the books, priceless artifacts from Strasbourg's French history, including the red bonnet placed atop the cathedral during the revolutionary Terror and Napoleonic general Jean-Baptiste Kléber's sword, as well as objects from its German history, including the original copies of Strasbourg's constitution, testimonies from the trial of Gutenberg, and copper plates for historic city maps, were all lost.[19] An end to the destruction would not come until 27 September, when a white flag appeared from the cathedral's single tower. Using legal jurisdictions passed during the French Revolution (the law of 8–10 July 1791 and the law of 23 May 1792), the German Army required the inhabitants of Strasbourg to provide food and lodging to the nearly eight thousand soldiers that had entered the city walls.[20] Shortly thereafter, the message came from German authorities: "From now on, Strasbourg is and will remain a German city."[21]

A commemorative German image of Strasbourg's capitulation captured the evolving role of a European border city as a symbolic war prize during a rising period of conflict-oriented modern nationalism (fig. 6.7). A compilation of different forms of visual media, the image assembled together a number of photographic views that provided direct, blunt evidence of a city humbled by the force of the German Army. With the exception of the centerpiece image of the cathedral and the portrait of Emperor Wilhelm, all of the views contained scenes of destruction. Constructing a visual narrative of militant-nationalist triumph, lithographic designs enveloped the city's

Figure 6.7 This image, commemorating Strasbourg's destruction during the siege, combined decorative lithography with photographic realism. *Gedenkblatt an die Capitulation Strassburgs, 27 Sept. 1870, durch Gen. Werder.* YC 9058 gr 1, Staatsbibliothek zu Berlin, Preussischer Kulturbesitz.

disfigured streets, buildings, and monuments (including the Rhine Railway Bridge in the bottom left corner) into the branches of a German oak tree, a symbol of the German nation. In addition to the allegory of the oak tree, the illustrator used lithographic supplements to portray Germania at the top of the image, a broken Napoleonic cannon under her feet, holding out the crown of victory in one hand and a sword in the other. While the image created the impression of an urban settlement brought to its knees by the German Army, it is notable that the people of Strasbourg were largely absent from the pictorial references. In choosing not to portray the suffering population in places like the city hospital, the image downplayed the psychological toll of the war on Alsatians whom the Germans viewed favorably as their long-lost brothers.

In contrast to the Germans' triumphal framing of Strasbourg's damaged cityscape, Charles Winter, one of France's most eminent nineteenth-century photographers, provided an alternative Alsatian perspective on the disfigured city in his album entitled *The Siege of Strasbourg*. The album's eighty photographs, silent witnesses to Strasbourg's devastation, offered a dark premonition of the total wars to envelop Europe in the twentieth century. One photograph of the destruction, taken from the observation platform of the cathedral, pictured a horrific version of the touristy "picturesque panorama."[22] Rather than capturing the familiar optimistic view of an endless, peaceful horizon, Winter's panorama of urban destruction underscored the fragility of European civilization in the face of warfare (fig. 6.8). Another photograph, taken from the Faubourg de Pierre the day after the capitulation, showed two German soldiers meditating on that fra-

Figure 6.8 Note how Winter's view from the cathedral inverts the notion of a "picturesque" urban panorama, offering instead a seemingly endless view of destruction. Charles Winter, *Belagerung von Strassburg*, 1871. Archives de Strasbourg, GF 148.

Figure 6.9 Two soldiers gaze across the wreckage of Strasbourg. See Charles Winter, *Belagerung von Strassburg*, 1871. Archives de Strasbourg, GF 148.

gility, gazing out toward the siege's path of destruction through the city (fig. 6.9). Viewing Winter's photographic album alongside map images such as the plan of shell strikes and civilian deaths at Strasbourg's municipal hospital, we see how nineteenth-century Alsatians used different kinds of popular visual media to communicate a message of victimization and tragedy in 1870.

"In Strasbourg," lamented one French writer, "the ruins themselves escape us: foreign hands will dress their wounds. . . . The burnt quarters emerge from ashes only to see a new master."[23] Indeed, the new government in Strasbourg did not wait long to turn its vision for a German Strasbourg into a reality. Already in the first week after the city's surrender, German authorities, in collaboration with municipal personnel, began to requisition workers to clean up the rubble.[24] Over the course of the next twenty-five years, the newly formed German imperial government would pay over one million marks to repair the damages caused by the siege.[25] Rather than serving as a mediating site for French-German exchange, Strasbourg's economic orientation would shift to the German markets to the east. But the memory of the siege would not fade quickly for the generation of Alsatians who had experienced it. Strasbourg emerged from the fires of 1870 as a city divided between soldiers and civilians, and Alsatians and Germans.

GRAND AMBITIONS: THE MAKING OF AN IMPERIAL
GERMAN BORDER CITY, 1871–1918

The annexation of Strasbourg in 1871 transformed the city into a German border space—now officially called Strassburg—that would play a key role in the development of a modern imperial German identity. As the western-most major urban outpost of the newly unified empire, Strasbourg became for German nationalists a symbolic site in which to define "Germanness." Strasbourg's changing cityscape from 1871 to 1918 can be viewed as a met-aphor for the complex political culture of an empire that was forced to maintain a delicate balance among the needs of the military, the bourgeois commercial class, the imperial state, and ordinary people. When construc-tion of the new Strasbourg was completed around the time of World War I, the formerly French provincial border city had been transformed into a major metropolis of southwest Germany. The Germanization of the city left its mark in numerous urban spaces: in the development of dozens of sophisticated military structures, in the establishment of imperial railroad and commercial networks, and in the construction of modern university, palace, and residential neighborhoods. Strasbourg would become one of the great success stories of the German imperial period, rendering visible the promise of the empire in its imposing fortresses, modern avenues, and grand architecture, as well as in its successful visual integration of the city's medieval German past with its modern German present.

Not unlike the militarized image of Strasbourg that nineteenth-century French officials had once promoted, German imperial authorities initially viewed their newly conquered border city as a fortress whose primary func-tion was to sustain a future war front against France. Along with the Lor-rainer city of Metz and a series of smaller garrison towns in the Vosges Mountains, Strasbourg played a critical role in Germany's plans to secure its new border. The development of Strasbourg's civilian neighborhoods would thus have to work around considerations of the city's defensive role. In 1871, the German Army reserved a ring of militarized territory (*Militär Gelände*), seven kilometers in radius, for a new military perimeter around the city. Shelling from the siege had destroyed much of the old city wall, built at the time of Louis XIV's military conquests, necessitating the con-struction of a new wall beginning in 1877. When completed, the wall was eleven kilometers long and twelve meters tall, composed of nineteen bas-tions with shooting positions accessible from ramps and equipped with cannons. The wall had a number of strategic openings, including eleven gates, one war gate (*Kreigsthor*, the only built in Europe from 1870 to 1888 and now classified as a historical monument), one war passage, and three tunnels for railroad tracks.[26] Forts were concentrated toward the west, in anticipation of a battle front against France. The German Reichstag provided three million marks to fund the wall's construction, money that in part came from French war reparations.[27]

Within Strasbourg, the German military constructed a number of new buildings that symbolized the empire's military strength vis-à-vis France, including the Kaiser-Wilhelm Garrison (situated within Vauban's citadel), the Manteuffel Garrison (the biggest in Strasbourg, housing three battalions and costing nearly two million marks), the Margarethen Garrison (housing ten companies from the Württemberg infantry), the Von Decker Garrison, and new barracks for the city's Esplanade.[28] A popular soldiers' song reflected the feel of Strasbourg as a German garrison town:

> Oh Strassburg, Oh Strassburg
> You most wonderfully beautiful city
> Inside lay hidden
> So many soldiers.
>
> So many a handsome
> And gallant soldier
> Whose father and mother
> Have abandoned him cruelly.
>
> Abandoned, abandoned
> It can be no other way!
> To Strassburg, yes to Strassburg
> Soldiers must go.[29]

To provide a structured social life for the troops, the city also allotted building space for a large military casino with eleven dining halls, six library rooms, social rooms, and a dance hall. In addition, the city constructed Protestant and Catholic garrison churches (seating twenty-one hundred and fourteen hundred persons, respectively) to keep watch over the soldiers' morals.[30] The visible presence of a large, rotating military population—including the regular appearance of a military band at the Rheinlust garden on Sunday afternoons—served as a reminder of the persistent state of tension between France and Germany. Particularly during visits from the German emperor, Strasbourg became an urban stage for the showy performance of German military power.[31]

But Germans did not simply view Strasbourg as a staging ground for wars; they also envisioned a future for the border city as a prosperous peacetime capital of imperial commerce and culture. Once the German military had claimed its ring of militarized land, Strasbourg's German-controlled municipal administration (the mayor and municipal counsel were appointed by imperial decree until 1886) initiated plans to triple the size of the city's civilian neighborhoods to accommodate the thousands of German immigrants that moved to Strasbourg to work as administrators, judges, railroad officials, university professors, and businessmen. Urban plans for the new, postsiege Strasbourg thus provide a rich source of visual evidence for the bright future that Germans imagined for their western border city.

In 1878, Johann Gottfried Conrath, the city's Parisian-trained chief architect who maintained his post after the annexation, publicly circulated his blueprints for Strasbourg's ambitious urban extension plan. To signal the German character of the city, Conrath conceived his spatial framework for Strasbourg's new districts in dialectic with Strasbourg's medieval town center, creating an interconnection between Strasbourg's German past and its German future through visual practices.[32] The key to Conrath's plan was his decision to use Strasbourg's medieval German cathedral as the orientation point for all of the new imperial German monuments in the city. He did so by establishing two major sight lines (see detail in fig. 6.10b) that originated at the corner of the cathedral in central Strasbourg and extended to the new imperial square (Kaiserplatz, now the Place de la République) and the military quarters (the Nicolaus-Kaserne and the Esplanade).[33] One commemorative image of the city's expansion process, picturing a medieval knight squashing the old city, captured Conrath's vision for a modern Strasbourg that was both forward and backward looking (fig. 6.11).[34]

Over the following decades, builders slowly filled in Conrath's blueprint for the expansion and modernization of Strasbourg with works of German architecture ranging from banks to courthouses to high-ceilinged apartment complexes. Many of the buildings exhibited a classical "Wilhelmine" style that members of the imperial elite favored as a universal architectural form that could symbolically unite new constructions across the empire.[35] The new imperial palace, for example, had an imposing, grandiose presence intended to symbolize the permanence of the German Empire in Alsace-Lorraine (see fig. 6.12 and note the connecting sight line with the cathedral). Collaborating with city architects and builders, members of Strasbourg's German immigrant community formed the Strasbourg Beautification Society (Verschönerungs-Verein Strassburg) to ensure that the new neighborhoods catered to bourgeois standards of taste and hygiene, with open air, parks, trees, and sunlight.[36]

The modern German city under construction in turn-of-the-century Strasbourg was designed not only for the viewing experience of residents, however, but also for the thousands of German tourists who visited Strasbourg each year. Many tourists were aided by guidebooks that interpreted the urban landscape for them, directing their gaze toward privileged spots. As one historian writes, Strasbourg's tourist guidebooks "inscribed the empire's mark into the urban landscape."[37] Emphasizing the past and present importance of Strasbourg to the German Empire, guidebooks typically split

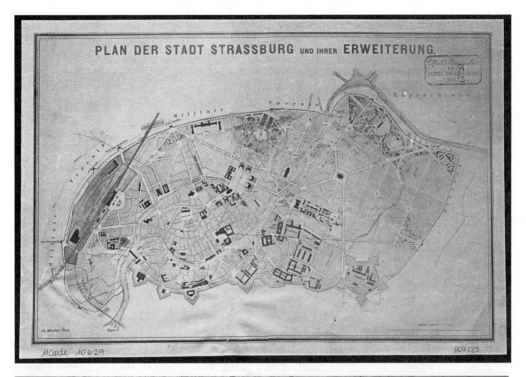

Figure 6.10a Conrath's new urban plan for Strasbourg was designed to connect the city's medieval cathedral to the new German imperial quarters through carefully planned sight lines. Johann Gottfried Conrath, *Plan der Stadt Strassburg und Ihrer Erweiterung*, 1878. Photo. et coll. BNU Strasbourg, M.Carte.10629.

Figure 6.10b Detail of the two sight lines connecting the medieval cathedral to the new districts of Strasbourg. Johann Gottfried Conrath, *Plan der Stadt Strassburg und Ihrer Erweiterung*, 1878. Photo. et coll. BNU Strasbourg, M.Carte.10629.

Figure 6.11 A dramatic illustration of Conrath's urban expansion plan for Strasbourg. *Kommers zu Ehren des Bürgermeisters der Stadt Strassburg Herrn Otto Back aus Anlass seines 70 jährigen Geburtstages veranstaltet von den städtischen Beamten am 20ten Oktober 1904.* Photo. et coll. BNU Strasbourg.

their pages between the Old City (including an obligatory visit to the top of the cathedral) and the New City (emphasizing the imperial palace). Some authors compared the promise of the city's German future to the failures of its French past. One guidebook explained: "From the humble departmental capital that was surrounded by an excessively tight fortified wall, with an insignificant existence that is characteristic of a French provincial city, Strasbourg became the capital city of a region [*Land*] whose rich resources found a lovingly nourished role in the empire."[38] The symbolic importance of Strasbourg to German national identity was a consistent theme. "No city's name," declared one guidebook, "arouses in the same manner every combination of melancholy and proud joy in the souls of the Germans as the name Strassburg."[39] Another text called Our Lady of Strasbourg "possibly the most prominent and imposing work of architecture that exists on German soil."[40]

However, the guidebooks that pictured Strasbourg through a German national framework glossed over the difficult day-to-day reality facing the city's residents: postsiege Strasbourg was a painfully divided city. Reflecting on Strasbourg's evolving late nineteenth-century cityscape, the president of the German Architectural Association proclaimed that the new buildings stood as "silent witnesses to the spirit that stands not only for the victorious battles fought, but also for the peaceful conquests to be made, hearts to be won."[41] Instead of winning over hearts and minds, however, Strasbourg's urban transformation had the immediate effect of dividing German and Alsatian social geography based on culture rather than class: the new German immigrant population, constituting 40% of the municipal population, settled in the New City districts, while rich and poor Alsatians alike concentrated in the Old City (*Altstadt* or *Kernstadt*).[42] Historians Georges Livet and François Rapp explain how the experience of the siege and the violent turn in French-German border relations put a strain on friendly encounters between Alsatians and Germans: "Up until 1900, the frontiers remained entrenched: two populations lived in Strasbourg juxtaposed. . . . Before 1870, relations were more open—the Alsatian and German bourgeoisie met in Baden; the social horizon of the great Strasbourg families extended to Basel, Frankfurt, and Cologne—while [afterward] the German and Alsatian bourgeoisie no longer mixed."[43]

KAISERPALAST   PALAIS IMPÉRIAL

Figure 6.12 The imperial palace (*Kaiserpalast*) was one of the major architectural projects that German immigrants undertook in late nineteenth-century Strasbourg. A Berlin-based architect designed the grandiose building to symbolize the German Empire's permanent hold on its border city. Image of the Kaiserpalast from Jul. Manias, *Album von Strassburg: 50 Tafeln in Lichtdruck nach Originalaufnahmen* (Strassburg: Jul. Manias, 1906). Photo. et coll. BNU Strasbourg.

Echoing the historians' observation in her autobiography, Elly Heuss-Knapp, who grew up in Strasbourg's New City as the daughter of a German professor at the new University of Strasbourg, provides a compelling narrative of the unofficial borders that divided the urban landscape. When Heuss-Knapp and her family moved to the New City, the area resembled a construction zone for new apartment buildings in the Wilhelmine style. According to Heuss-Knapp, everyone in her neighborhood was German: fathers had come from Germany to work as professors, functionaries, or officers. "Even though we were young," she recalls, "we only spoke with the neighboring German children; the others were a world apart, interesting but inaccessible."[44] Later in life, Heuss-Knapp noted the differences between Strasbourg's elite circles of Alsatians and Germans, writing: "The Alsatian upper class carried out a social life that was almost completely cut off from that of the Germans. . . . I was always full of astonishment to see how, in a modestly sized city such as Strasbourg, groups could go about an existence that was so separated."[45] Engaging in visual politics themselves, Alsatians would attempt to denationalize Strasbourg's cityscape and frame their hometown as primarily a *regional* capital whose appearance was neither French nor German. Two visions of Strasbourg therefore emerged at the turn of the century: that of German immigrants and tourists, who viewed the urban cityscape from a German imperial perspective, and that of disoriented Alsatians, searching for an urban identity outside the spatial framework of empire.

NOSTALGIC CITYSCAPES AND THE REJECTION OF NATIONAL ORIENTATIONS

In his humorous guidebook, published in 1883, Rodolphe Reuss, a prominent journalist, librarian, and member of an old Protestant Strasbourgeois family, described the city's German-led expansion from an Alsatian perspective. Referring to himself as a "flâneur" and "a simple bourgeois from Strasbourg," Reuss takes his readers on tour of the city's recently built districts and their imposing streets bearing names from Alsace's Germanic past. To help him navigate the new spaces, Reuss purchases a copy of Conrath's 1878 urban plan of Strasbourg's New City, an area that he terms an "immense desert" of "unknown regions" that stands to become "the promised land of future generations."[46] As he walks through the new districts, Reuss makes fun of the Germans' choice of street names. Why, he asks, is a new quay named Kochstaden after the cooking profession? Why choose to name one of the city's new squares after Sebastian Brant, a satirist famous for writing about a "ship of fools" (*nef des fous*)? And what of the French place-names that had vanished without a trace? Reuss looks through the address book for the city square named after the French revolutionary Benjamin Zix only to find that the name no longer exists. In spite of its lighthearted tone, Reuss's text was a powerful commentary on the disorientation of Strasbourg's Alsatian community after the German annexation. Adjusting to life under German rule, Strasbourg's residents were confronted with the tenuous nature of their national identity: within the span of a few years, they had witnessed the official obliteration of the city's French past and the rediscovery of its German one.

Visual images of Strasbourg's lost cityscape expressed the Alsatians' dissatisfaction with the German-led modernization of their city and their nostalgia for their small provincial city. Focusing on historic views of Strasbourg from old books and map images, Adolphe Seyboth's *Views of Old Strasbourg*, published in 1891, was popular among the city's Alsatian bourgeoisie. The beautiful picture album reflected a sense of mourning for Strasbourg's Old City that was not simply antiquarian, but also political in nature. The book's stated goal was to "forever keep alive" the historical memory of Strasbourg's lost cityscape, noting that "the medieval character of the city has almost completely disappeared."[47] Seyboth's text reads as both a skilled work of archival history and a commentary on Alsatian regionalism. In grieving the disfiguration of old Strasbourg, he critiques not only the recent German constructions, but also several prominent eighteenth-century French edifices, including the Rohan Palace, the Town Hall, and the Aubette Theater. Focusing his attention on the "authentic" parts of the city, Seyboth raises to iconic status buildings from neighborhoods that upstanding Alsatians had only recently been afraid to walk through. The working-class fishmonger district of Petite France, for example, acquires a regionalist cachet, finding a prominent place in Seyboth's text as the site of Strasbourg's oldest half-timbered houses (*Fachwerkhäuser*, or *maisons à colombages*) whose building materials and facade design were unique to the Alsatian region. Rather than categorize the houses as French or German, Seyboth simply refers to them as "Alsatian."

By the early twentieth century, the Alsatian public's nostalgic feelings for old Strasbourg had led to an organized movement to preserve the city's historic buildings from destruction. At this point, Strasbourg's building commission had begun to raze the city's historic neighborhoods in order to build wide avenues for shopping and commerce, a long-term project called the "great street clearing," or *grosser Strassendurchbruch*. Pressure mounted in Alsatian civil society to protect the buildings that embodied Strasbourg's "local image." A sign of rising frustrations, Charles Winter's photographs of old Strasbourg taken after the siege of 1871 were posthumously reprinted in 1909 under the nostalgic title *Strasbourg disparu*. Warning that "we cannot and must not fall into the spokes of the grinding wheel of time," a local professor formulated a plan to establish legal oversight of new build-

Figure 6.13 Alsatian regionalists preferred to visualize Strasbourg as what it once was: a provincial cityscape rooted in local Alsatian culture. Note the use of the horizontal view in this image, which provides the viewer with a more intimate, human-scale view of the city. Adolphe Seyboth, *Ansichten des Alten Strassburg: Fünfzig Tafeln mit erklärendem Text* (Strassburg: J. H. Heitz, 1891). Photo. et coll. BNU Strasbourg.

ing constructions.[48] "The formation of the local image," Professor Emerich wrote, "which the population constantly has before its eyes from the earliest childhood onward, makes an impression that can last for an entire lifetime."[49] In 1910, the city responded to public pressures and passed a law restricting construction practices in historic districts.[50] A supplement to the law restricted the use of decorative paintings, signs, inscriptions, windows, gardens, or any "modifications that would be judged by the mayor as constituting a transformation of a certain importance," adding that building authorization would be refused if "the construction or modification does not give an agreeable impression to the eye or if it is of such a nature that it alters completely or part of the physiognomy of the areas."[51]

Paradoxically, the Alsatian-led protests against the Germanization of Strasbourg's cityscape were tied to a broader, pan-German *Heimat* movement that rejected a uniform imperial architectural style. Historian Annette Maas attributes the Alsatians' successful protest against German-led imperial urbanization precisely to their participation in an inner-German discussion of *Heimat*.[52] Across Germany, associations such as the Bund für Heimatschutz were challenging an older model of German unification that advocated the use of "national" architectural styles to create an imperial identity. The concept of the "local image" that Alsatian regionalists adopted was part of a new approach to German urban planning that celebrated the empire's cultural diversity.[53] According to this new nation-building strategy, cities could appear "German" because of their organically developed architecture based on local building materials and "authentic" styles. It was a decentralizing aesthetic trend that offered a promising path toward fuller Alsatian participation in the German national project. World War I, however, prevented the continuation of this trend, abruptly thrusting the question of Strasbourg's regional identity into French hands. During the turn of the century, French citizens had also attempted to balance the pressures of modernization with their desire to protect their regional patrimony, passing laws that restricted new buildings in favor of historical preservation in 1887 and 1913. But France's initial postwar impulse toward urban planning in Strasbourg would be similar to that of the Germans' after the siege of 1870. The French would move quickly to put their own visible mark on Strasbourg's cityscape.

THE POLITICS OF SEEING AND FORGETTING IN POSTWAR STRASBOURG

The entry of French troops into Strasbourg on 22 November 1918 was one of the most impressive visual performances of French military power in the postwar period. Coordinating their parade schedule with the mayor days in advance, Marshal Foch and his troops were greeted by the sound of the city's booming cathedral bell and numerous smaller church bells.[54] The municipal council put away the red flags from the short-lived soldier and workers' commune, flying tricolor flags throughout the city instead. Cheered by enormous crowds, the troops marched through the city and paid homage to prominent sites of French national memory, including the statue of the Napoleonic hero General Kléber. But France's symbolic national conquest of Strasbourg also involved a military performance at a prominent symbol of German imperialism: Emperor Wilhelm's imperial palace. Rather than avoid the iconic German building, the French military chose to invert its meaning by showing their domination over it. Municipal authorities helped to choreograph the French takeover of the symbolic site of German nationalism, unlocking the doors and opening all the windows of the palace in advance of the French arrival to stage a dramatic photograph.[55] The image of jubilant French troops bursting out of the windows and porches of the monument to German imperial power

Figure 6.14 In staging one of the parade route's ceremonies at Emperor Wilhelm's former imperial palace, the French military displayed its triumph over German imperialism in Alsace-Lorraine. See the postcard *Le général Gouraud passant en revue les troupes françaises à Strasbourg, le 22 novembre 1918* (Strasbourg: Maison d'Art alsacien, 1918). Photo. et coll. BNU Strasbourg.

became an iconic symbol of the enemy's "excommunication" from the border city.[56]

In the years to follow, the French continued to manipulate Strasbourg's public image to make it appear less German. Faced with an urban plan of Strasbourg that was already filled in by turn-of-the-century German architectural styles, however, proponents of the city's "Frenchification" had to rely primarily on written narratives and visual framing techniques to make the argument that Strasbourg was indeed an authentically French city. The postwar street-naming commission, for example, set to work on its efforts to rename, or *débaptiser*, the city's German-sounding toponyms.[57] The commission received numerous suggestions from an Alsatian public desiring to "erase the past from our view" and to eliminate German names that were "reminiscent of the old traditions of a district made up entirely of *boches*."[58] Residents of one city block, for example, signed a petition to change their street name to Avenue Clemenceau.[59] French guidebooks to the city also attempted to turn public taste against Strasbourg's modern German buildings, criticizing them as "propaganda by stone" in the spirit of German *"Kultur."*[60] Michelin's 1921 guide referred to the city's German constructions derisively as "vast" and "colossal," directing visitors instead to buildings in the French or provincial style such as the Rohan Palace, Petite France, the Places Broglie and Kléber, and the monument to General Louis Desaix, who had defended the passage to Kehl against the Austrians during the revolutionary war.[61] In addition to the guidebooks, French transitional authorities supported the construction of a new city museum that reinforced a narrative of Strasbourg's urban spaces that was deeply rooted in French national history and tradition. Frustrations with the city's appearance, however, sometimes still rose to the surface. "Down with the German *Ortsbild!*," read a 1926 editorial in one French-language newspaper. "Our city must acquire a French cachet."[62]

In addition to the competition between French and German national architectural styles in Strasbourg, Alsatian regionalists continued to make efforts to preserve the city's unique provincial appearance. Organizations in civil society, such as Les Amis de Strasbourg, advocated for the preservation of the city's historic buildings. The municipal law of 29 July 1925 reaffirmed the "protection of local image" laws passed under German rule. In addition, the city provided financial support on a case-by-case basis to help owners clean and restore the facades of iconic, Alsatian-style half-timbered houses classified as historical monuments.[63] At times, Strasbourg's Alsatian-led city government even took stands against creating a national French appearance for the city. By the mid-1920s, for example, Mayors Charles Hueber and Jacques Peirotes had changed a number of street names from French to bilingual French and German forms. Strasbourg's new buildings, modest in number within the central city, reinforced an interwar compromise of regional, modernist, and French styles.[64]

VISUALIZING STRASBOURG AS A FUTURE PORT CITY

During the interwar years, questions about the future of Strasbourg's cityscape focused not only on its commercial and residential neighborhoods, but also on the development of its port. A French map from 1919, printed on top of an earlier German map of the city, helped French urban planners to see the deep interconnections between Strasbourg's urban topography and the canalized shape of the Rhine River.[65] What this particular map did not reveal was the extensive national security regime that the Treaty of Versailles imposed on the Rhine border zone during the same year. While the treaty ostensibly gave the bridges that traversed the Rhine "shared" national status, it granted France sole control over the railroad lines and telephone wires that crossed the bridges and permitted the French to maintain facilities on the river's right bank, in German territory. It also gave the French control over the German port city of Kehl, located directly

Figure 6.15 Strasbourg's urban topography was deeply interconnected with the Rhine River. When tensions between France and Germany grew during the interwar years, it became increasingly difficult to cross the city's network of Rhine bridges. Service cadastrale de la ville de Strasbourg, *Carte des ports du Rhin de Strasbourg et de Kehl*, 1919. Archives nationales, 30 AJ 213.

across the Rhine from Strasbourg, for the first seven years after the armistice. The line of sovereignty between the two countries, moreover, had to be marked "in a very clear manner" on a bilingual table at the midpoints of the bridges.[66] The Treaty of Versailles thus promoted a view of Strasbourg's Rhine topography as a zone where the French-German boundary needed to be strictly monitored and enforced.

The postwar vision of the Rhine border as a French national security perimeter was challenged, however, by French business leaders who wanted to promote open commercial exchange with German and other European markets. In 1924, Strasbourg opened a new port designed to benefit from the city's advantageous position along the great international waterway. Blueprints for the new port reveal how it expanded significantly beyond the existing German-built port from the turn of the century. Shortly after the port opened, entrepreneurs began construction on coke and thermal factories. The French government also helped to fund a brand-new Rhine fleet to encourage cross-border economic exchange with other river nations. But perhaps the boldest Rhine project undertaken on the outskirts of Strasbourg was the Great Alsatian Canal, begun in 1921 to solve the problem of Rhine navigation from Basel, Switzerland, to Strasbourg. Designed by Alsatian engineer René Koechlin, the cement canal was to run parallel to the Rhine, diverting the water from its original riverbed into massive man-made structures. Eight hydroelectric stations were to be built along the canal, generating energy from a series of strategically placed artificial waterfalls. One of the largest rectification projects undertaken on the Upper Rhine since the Tulla Rectification Projects (1817–76), the plan received funding from the French government, local Alsatian banks, and the Société Énergie Électrique du Rhin.

French members of the newly founded Rhine Commission, housed in Strasbourg at the German emperor's abandoned imperial palace (subsequently called the Palais du Rhin), tried to market the Great Alsatian Canal to the general European public at the International Day of Domestic Navigation in 1926. The meeting, which was held in Basel, brought together Rhine organizations from a number of European states. It was at this exhibition that Gaston Haelling, the director of the port of Strasbourg, delivered a passionate defense of France's decision to construct the Great Alsatian Canal: "Is it not a beautiful act of faith in human agreement, in the energy deployed to use all the resources that Rhine navigation gains for France, an act of faith in the brotherhood of peoples, an act of faith in the future peace, an act of faith that not only in France will be thought of one day as an act of boldness?"[67] Without categorizing people according to their ethnicity, religion, or culture, Haelling imagined a day when Europeans could peaceably use the natural resources laid before them to create a lasting peace. The Rhine constructions reflected the hope that the French-German borderland could become a site of transnational exchange, resurrecting the spirit of mid-nineteenth-century liberalism that had encouraged an open and optimistic view of Europe's national borders.

Director Haelling's vision of Strasbourg as a bustling international port echoed some of the more theoretical discussions about the meaning and function of national boundaries that took place simultaneously at the city's renowned university. During the interwar years, the University of Strasbourg was the intellectual home of some of the most innovative and influential scholars in twentieth-century France, including Marc Bloch and Lucien Febvre. Emphasizing the power of geography and long-term social change to shape history, these scholars, whose collective work became known as the Annales school, took a stand against the "natural borders" theory that had shaped French international relations during the eighteenth and nineteenth centuries. Lucien Febvre wrote in 1924: "Mountains, rivers and forests . . . are undoubtedly boundaries in so far as they are often real obstacles. But they are also bridges, centers of expansion and radiation, little worlds with attractive values of their own, linking together the men and the regions on either side of them. . . . They are never boundaries 'of necessity.'"[68] Rather than interpret the Rhine as a mechanism of separation between peoples, these interwar historians and geographers from the University of Strasbourg preferred to think of the Rhine as a source of union.

The Annales school's alternative view of the Rhine as a unifying geographic feature was soon overwhelmed, however, by rising border suspicions and tensions between the French and the Germans during the 1930s. Keeping to the agreements established by the Treaty of Versailles, the French Army pulled out of its occupation zone in the Rhineland in 1930. Germans living in the city of Kehl publicly celebrated this much-anticipated departure of French troops. Later, in 1934 and 1935, residents in Kehl could be heard firing cannon shots celebrating the anniversary of Germany's victory over France at Sedan in 1870.[69] Noting the increased presence of SS and SA officers at the customs offices on the other side of the Rhine, French police in Strasbourg closely monitored border crossings and restricted the availability of tourist passes.[70] Alsatian police kept a file on the cross-border travels of Hans Luthmer, the mayor of Kehl who had grown up in Alsace and was known to support Germany's reclaiming of the border region.[71] Adding to the tensions, the crash of world financial markets during the 1930s had a severe impact on the construction work along the Rhine. Banks that were once sufficiently confident to invest in the Great Alsatian Canal and the port of Strasbourg chose to invest their money elsewhere, and work halted on both projects, delaying the canal's completion until 1959. In a last symbol of the broken state of French-German relations, retreating French troops dynamited the beautiful nineteenth-century Rhine River

Bridge connecting Strasbourg and Kehl in 1940, shortly before the Germans reclaimed the city for a second time.

Using maps as their primary thinking tools, French and German urban planners conceived of new ways to orient Strasbourg's streets, waterways, and public spaces in order to demonstrate the city's political and cultural links to their nation. Over time, the top-down views of Strasbourg pictured on their urban plans resulted in tremendous changes to the city's physical makeup. The idealized sight lines that urban planners drew on paper led to massive street-clearing projects that demolished parts of old quarters in favor of broad avenues. They made possible the construction of new landmark buildings that argued for Strasbourg's place within the French or German sphere of influence. But evidence from Strasbourg's map archive also reveals that the French and Germans were at times willing to consider an alternative vision for their border city. Some innovative urban planners preferred to imagine a future for Strasbourg as a bridge, rather than a point of rupture, between two powerful European civilizations. It is this vision of Strasbourg as a "European" capital city that shapes much of its urban development today.

Several years ago, when I was walking through a Paris metro station on a cold winter day, I noticed an advertisement for an exhibition on historical relief maps of France. The exhibition, called *La France en relief*, was set to close in a few days, so I changed metro lines and headed straight to the Grand Palais, a grandiose Beaux-Arts-style structure situated along the Seine. When I entered the cavernous building with its looming glass roof, an incredible scene unfolded before my eyes. Spread across the vast exhibition hall were sixteen enormous relief maps, some as large as 160 square meters, representing different cities across France. From the time of Louis XIV until the late nineteenth century, the maps served as visual aids for military leaders charged with the defense of French cities, particularly border cities. Encased in protective glass, the maps were artfully lit and accompanied with plaques that explained different aspects of the city plans. As I walked from map to map, I felt as though I were taking part in a historic *tour de France*. The maps transported me back to a time when a militarized vision of French territory was still dominant, and the European competition for border cities like Strasbourg remained fierce.

While the enormous maps on display at the Grand Palais were fascinating in themselves, it was just as intriguing for me to observe the crowds of people viewing the maps. The exhibition hall overflowed with French people of all ages. Children ran excitedly from one replica of a French city to another, as if the relief maps formed a sort of interactive playground. Elderly men with wool hats and long coats huddled in small groups on the observation platforms that surrounded the maps. Their faces became animated as they pointed to the miniature replicas of the ramparts, canals, houses, cathedrals, and streets that once made up France's major cities. Near the exit, a line of people waited to have their photos taken in front of a giant map of France.

Put into historical perspective, the bustling scene at the Grand Palais revealed the enduring legacy of a cartophilia—a popular love of maps—that spread across France during the previous two centuries. The image of a hall packed with ordinary Parisians, each paying a nominal fee of 2.50 or 5 Euros, may well have been a scene from the Paris Exhibition of Geographical Sciences in 1875, where a similar demographic of regular French men and women, as well as children from nearby schools, came to see the cutting-edge work of mapmakers from across Europe. The difference, of course, is that visitors at the recent Grand Palais exhibition saw the maps surrounding them as visual artifacts from an older, now defunct, territorial order. The exhibition's purpose was to serve as a window into the mindset of another time, when Europeans were intent on using maps to visualize and enforce the boundaries that separated them. If the enormous military relief maps on display at the Grand Palais offered a glimpse backward into Europe's territorial past, then what steps are Europeans taking to plan their territorial future?

The focus of European cartography today is far different from what it was during the nineteenth and early twentieth centuries. In the past, European governments and civil associations expended a great deal of their time and financial resources on making national boundaries appear more visible, both on paper maps and on the physical landscape. Since the end of World War II, however, the momentum has swung in the other direction. Supporters of European integration have striven to make national borders between their countries *less* visible. In an effort to build a lasting peace between former enemies, their goal has been to transform Europe into an unbroken canvas of land, free of national ruptures, where people and goods can travel peacefully. It is a bold vision of European geography that has in some ways achieved remarkable success. Today, passengers on a high-speed train from Paris to Berlin will look out their windows and see a French-German border that appears nothing like it did one hundred years ago. Symbols of the tense border relationship of the past—the watchtowers, armed guards, and customs stations—are gone, replaced by gas stations and chain stores. This new border reality stands as a testament to the leadership of European visionaries like Jean Monnet and Robert Schuman, two men who experienced firsthand the passions of nationalist hatred in France and Germany and saw an integrated European geography as a means to end the cycle of violence between their countries.

There is no better place to witness the postwar transformation of European border space than the Alsatian city of Strasbourg. Called the "Capital of Europe," Strasbourg has become a showcase for the kind of intergovernmental and cross-cultural cooperation that European integrationists desire. In recent decades, a new "European Quarter" has risen up alongside the *Neustadt*, or New City, that German imperialists built on the outskirts of old Strasbourg during the late nineteenth century. The European Quarter is home to numerous institutions that anchor the European integration movement, including the European Parliament, the Council of Europe, and the European Court of Human Rights. Strasbourg also serves as the headquarters of cross-border cultural institutions such as the public television station ARTE, which offers bilingual programming in French and German. Farther afield, on the eastern edge of the city, the recently constructed Bridge of Europe links Strasbourg to Kehl, the German city located just across the Rhine. Replacing the Gothic-style Rhine Railway Bridge that French troops dynamited during World War II, the new bridge, completed in 2004, is built with modern, flexible materials that are bent to appear as two hands extending from either side of the border into a handshake.[1]

But creating a more integrated European geography is not just a matter of erasing long-standing national boundaries. It is also about constructing, and rendering visible, new kinds of bounded territorial units that reflect a transnational European identity. Over the past several decades, proponents of European integration have begun to map and nurture new spatial units called Euro-regions.[2] In terms of their size, Euro-regions are analogous to the historic provinces of Europe's past, such as Alsace and Lorraine. What makes Euro-regions so different, however, is that they are composed of small chunks of territory from three or four different countries. Southern Alsace, for example, has joined Baden and northwest Switzerland to form the futuristic sounding Euro-region of TriRhena.[3] Lorraine, meanwhile, has joined Luxembourg, parts of the German Saarland, and the Walloon part of Belgium to form the Euro-region Saar-Lor-Lux. The goal of these transnational border zones is to promote economic vitality through the free exchange of labor and goods. But how successful have the Euro-regions been in shifting popular mentalities toward national borders? Do people actually feel attached to their Euro-region?

The hard reality is that only a small group of Europeans are aware of the presence of Euro-regions in their daily lives. Most of the people who create and look at maps of Euro-regions work for the alphabet soup of impersonal European institutions such as ESPON (the European Spatial Planning Observation Network) or CORINE (the Coordination of Information on the Environment).[4] With their circulation limited to the technocratic sphere, maps of Euro-regions remain strange and unfamiliar to most people; they have yet to become the kind of "logo maps" that enabled the visual form of the nation-state to seep into minds of ordinary people a century ago.[5]

The slow popular acceptance of the Euro-region concept demonstrates that making maps is just the first step toward creating a new territorial order. Advocates for reform must also teach the public to believe in the cultural and political legitimacy of the new territorial spaces that the maps represent.

It follows that one of the greatest stumbling blocks to the public acceptance of Euro-regions is the entrenched power of national education systems. While the 1992 Maastricht Treaty encouraged European Union member states to develop a "European dimension" to their educational policies, the agreement also preserved each European country's right to make its own decisions about school curriculum.[6] This is why European students, for the most part, continue to learn a nation-based geography curriculum. Most of the maps that European students see in schools are the "logo maps" of national territory that date back to nineteenth-century classrooms. Nation-based geography education, moreover, is bolstered by institutional support from national mapping bureaus like the Institut national de l'information géographique et forestière (IGN) in France and the Bundesamt für Kartographie und Geodäsie (BKG) in Germany. For the growing numbers of Euroskeptics who crave a return to the age of national sovereignty, the persistence of national geography curriculums is clearly playing in their favor.

But are the prospects of creating a more unified European geography really so slim? This study has argued that the key to transforming popular mindsets toward European boundaries is the involvement of citizen mapmakers. If we focus on the current initiatives of European civil associations, we see that there are indeed some groups of citizens that are working successfully to dislodge the nation-state framework from the map of Europe. One of the most impressive examples of bottom-up mapping initiatives comes from the European Cyclists' Federation, an organization that represents forty countries and has more than five hundred thousand active members.[7] In recent years, the federation has mapped out fourteen long-distance bicycle rides across multiple European countries, including one route (no. 15) that follows the course of the Rhine River from its origins in Switzerland to the mouth of the sea in the Netherlands. The cycling federation hopes to complete construction of its network of "EuroVelo" bike routes, covering a total of seventy thousand kilometers, by 2020. In addition to its map of bike routes, both existing and projected ones, the federation has produced a guide to European cultural sites, ranging from war monuments to archaeological sites, for prospective bike riders to explore along their cross-border journey. It is this sort of grassroots mapping initiative, involving mapmakers and map readers outside government institutions, that has the potential to ground a transnational European identity in territorial reality.

At this time, it is still too soon to predict whether European boundaries will become more or less visible, or whether a new organizing structure for European territory, such as the Euro-region, will one day replace the nation-state. What we do know for sure is that maps will continue to shape how Europeans identify themselves and how they relate to their neighbors. Every generation of Europeans brings new kinds of mapmakers who invent new genres of maps to answer the key territorial questions of their day. In the near future, it will probably be digital maps, ephemeral images visible on computer screens for only days or weeks at a time, that will shape how Europeans see and interpret the borders that they share.

This book was made possible through the support of many generous people. First, I would like to thank two exceptional teachers who have shown me, through their dedication and joie de vivre, how to "be in the world" as a historian. John Merriman's passion for the land and the people of France, particularly for ordinary people, has been an inspiration. When I was a graduate student, John would always remind me that I was not writing about maps. I was writing, first and foremost, about the people who used maps. Not only is John a talented French historian, he is a great person. His generosity and kindness over the years, both in the United States and in France, has been humbling, and I am incredibly grateful for his friendship. I would also like to thank Mary Louise Roberts for showing me, early on in life, what it means to be a historian. I still remember sitting in a cavernous auditorium during my freshman year of college and being riveted by Lou's lectures on European civilization. I couldn't believe my luck when Lou, that charismatic figure from the lecture hall, agreed to work with me on my senior thesis, even helping me to plan my first archival research trip to France. In the years since I graduated college, Lou has continued to be a trusted friend and teacher. She read every word of this manuscript, and her unwavering enthusiasm made the completion of this book possible.

In addition to wonderful teachers, I owe a great deal to the institutions and foundations that provided me with financial support, particularly during the early stages of my map research in graduate school. I am particularly grateful for a Georges Lurcy Fellowship that funded me through twelve months of archival research in France, and for a Robert Leylan Fellowship for financing a year of writing after I returned from abroad. I would also like to thank the Yale Council for International and Area Studies (now the

MacMillan Center) for supporting two summers of research in Europe, as well as the Woodrow Wilson National Fellowship Foundation for providing me with a Mellon Fellowship to attend graduate school. At Montana State, a Scholarship and Creativity Grant provided vital travel and writing support for revising my manuscript. For their generous assistance in financing the book's sizable art program, I would like to thank David Cherry, chair of the Department of History and Philosophy, and Nicol Rae, dean of the College of Letters and Science at Montana State University.

I am grateful to numerous faculty members from the Department of History at Yale University for their help in conceptualizing this project. Ute Frevert, Laura Engelstein, and Bruno Cabanes all read an earlier version of this manuscript, and their insightful comments helped to make this book better. For their help along the way, I would also like to thank Paul Freedman, Chuck Walton, Bob Harms, Tim Snyder, and Steve Pincus. In addition to support from the faculty, it was my privilege to be surrounded in New Haven by an exceptional circle of graduate students in French history: Jennifer Boittin, Chris Brouwer, Kate Cambor, Rachel Chrastil, Katherine Foshko, Charles Keith, Ken Loiselle, Sarah Spinner, and George Trumbull. For their friendship and occasional runs to Modern Pizza, I would also like to thank Daniel Brückenhaus, Sarah Cameron, Faith Hillis, Anja Manthey, and Charlotte Walker-Said.

One of the best things about the history of cartography is that it attracts a diverse group of interdisciplinary scholars specializing in all parts of the globe. In recent years, I have been welcomed into a warm and friendly map history community that has greatly enriched this book. In particular, I have benefited from conversations with Jim Akerman, Jean-Marc Besse, Martin Brückner, Tony Campbell, Ray Craib, Catherine Delano-Smith, Matthew Edney, Morgane Labbé, Carla Lois, Bernardo Michael, Gilles Palsky, Neil Safier, Susan Schulten, Bernhard Struck, and Bill Wyckoff. I would also like to extend a special thanks to the participants who came from far and wide to attend a conference that my colleague Billy Smith and I organized, entitled "Mapping History," at the 320 Ranch in Big Sky, Montana. It was amazing to see what horseback riding, fly-fishing, an early October snowfall, and a lack of cell phone coverage did for our thinking about maps.

Almost all of the archival research for *Cartophilia* took place in Europe, and this book would not have been possible without the generosity of a wonderful group of colleagues, archivists, and librarians in France and Germany. In Strasbourg, I found a remarkable group of people eager to support my research on maps. Isabelle Laboulais at the University of Strasbourg, a renowned expert on the history of cartography, has been a fantastic colleague over the years, even helping me to secure a visa. Agathe Bischoff-Morales at the Médiathèques de la Ville et de la Communauté urbaine de Strasbourg welcomed me with a big smile to her library's rare book collection and generously provided me with office space. Daniel Bornemann and the staff at the Bibliothèque nationale et universitaire de Strasbourg paged hundreds of maps for me and later helped me to digitize many of them. Adélaïde Zeyer and the staff of the Archives départementales du Bas-Rhin never failed to be friendly and helpful, encouraging me early on to research the history of Alsace. Markus Heinz at the Staatsbibliothek zu Berlin generously helped me to locate the maps that I needed. At the Bibliothèque nationale in Paris, Catherine Hofmann offered invaluable assistance at key moments.

During my frequent research trips in Europe, I relied on numerous people to host me in their homes, keep me company, and explore. In Strasbourg, Yasmin Bari and Corin Warden made my research year abroad one of the best of my life. The Flammenküche always tasted better with them. For their hospitality and generosity over the years in Paris, I thank Annick Beillet, Sarah Griswold, Estelle Halevi, Michael and Maureen McMurphy, Evan Spritzer, Jonathan Szajman, Raphael Tyszblat, Jakob Vogel, and Robert Walker. In Berlin, Vera Pindter was great company and offered me a place to stay numerous times. Zoë Chafe and K. C. George experienced all corners of Europe with me, by bicycle, train, car, or whatever form of transportation that it took. Thank you for making my "work" on European history such a pleasure.

In the United States, my colleagues in the Department of History and Philosophy at Montana State University have been nothing but incredible. I am fortunate to teach in such a kind, generous, and good-humored humanities department that is a model of successful interdisciplinary collaboration. David Cherry, department chair, has provided moral support for this project from the beginning. Cassandra Balent is a brilliant manager and a visionary leader. I owe her so much for helping me to navigate through mountains of paperwork. I owe a special thanks to my senior colleague Billy Smith for collaborating on our memorable ranch conference. I also want to thank my colleagues Prasanta Bandyopadhyay, Rob Campbell, Susan Cohen, Dan Flory, Maggie Greene, Robin Hardy, Kristen Intemann, David Large, Tim LeCain, Sandy Levy, Dale Martin, Michelle Maskiell, Jim Meyer, Mary Murphy, Michael Reidy, Bob Rydell, Bart Scott, Molly Todd, Brett Walker, and Sara Waller. I also owe a big thanks to David Agruss, Laura Burkle, Jamie McEvoy, Jessica Marks, and Megan Higgins for their friendships in Big Sky country.

I am grateful to the University of Chicago Press for bringing this book to fruition. Abby Collier shepherded the project through the press at its early stages. Mary Laur and Logan Ryan Smith guided the book through production with care and good humor. Erik Carlson did an excellent job of improving the book's language. Kristen Raddatz was a great help with promotions. For their detailed comments on the manuscript, I would like to thank Ray Craib, Steve Harp, and a third anonymous reader. I would also like to thank Michele Angel, a talented artist and cartographer, for drawing the beautiful reference maps for this book.

Last, I would like to thank my fantastic group of friends and family from the San Francisco Bay Area. My friends Bríd Arthur, Keira Goldstein, and Sonali Bhagat have been steadfast and wonderful. A huge thanks to my big extended family: Mini, Fr. Peter, Lizzie, Grish, Serge, Fr. John, Bea, Vania, Jacob, Cookie, Nick, Sophia, PiePie, Tom, Tata, and Sasha. Finally, I would like to thank my parents, John Barrett Dunlop and Olga Verhovskoy Dunlop, who celebrate their fiftieth wedding anniversary this year. They taught me a lasting respect and admiration for foreign cultures and languages. Their love and support has been unconditional. I dedicate this book to them.

Introduction

1. On the rise of mass-produced visual images in modern Europe, see Walter Benjamin, "The Work of Art in the Age of Mechanical Reproduction," in *Illuminations*, trans. Harry Zohn (London: Fontana, 1968), 214–18. See also Bruno Latour's discussion of "immutable mobiles"—fixed images with a wide geographic circulation—in Bruno Latour, "Visualization and Cognition: Drawing Things Together," *Knowledge and Society: Studies in the Sociology of Culture and Present* 6 (1986): 1–40.

2. The term "counter-mapping" refers to mapping initiatives generated in reaction to the dominant cartographic paradigms of their time. Nancy Lee Peluso first used the term in a discussion of indigenous resistance to forest administration authorities in Indonesia. See Nancy Lee Peluso, "Whose Woods Are These? Counter-mapping Forest Territories in Kalimantan, Indonesia," *Antipode* 27, no. 4 (October 1995): 383–406.

3. The most recent comprehensive studies of Alsatian maps were published over fifty years ago. See Franz Grenacher, "Current Knowledge of Alsatian Cartography," *Imago Mundi* 18 (1964): 60–77; François de Dainville, "L'Alsace comme la voyaient les cartes anciennes," *Saisons d'Alsace* 22 (1967): 153–76.

4. On the role of borders in modern political culture and identity politics, see Malcolm Anderson, *Frontiers: Territory and State Formation in the Modern World* (Oxford: Oxford University Press, 1996); Hastings Donan and Thomas M. Wilson, eds., *Border Identities: Nation and State at International Frontiers* (Cambridge: Cambridge University Press, 1998); Richard Robinson, *Narratives of the European Border: A History of Nowhere* (New York: Palgrave Mac-Millan, 2007).

5. In his influential book, Thongchai Winichakul argues that national maps were often oriented toward the future. "A map," he writes, "anticipated a spatial reality, not vice versa. . . . A map was a model for, rather than a model of, what it purported to represent." See Thongchai Winichakul, *Siam Mapped: A History of the Geo-body of a Nation* (Honolulu: University of Hawaii Press, 1997), 130.

6. Christian Jacob has illustrated how maps were masterly mediums of communication that commanded great persuasive and rhetorical powers. See Christian Jacob, *The Sovereign Map: Theoretical Approaches in Cartography throughout History*, trans. Tom Conley (Chicago: University of Chicago Press, 2006), xiii–xv. See also Jeremy Black, *Maps and Politics*

(London: Reaktion Books, 1997); Mark Monmonier, *How to Lie with Maps* (Chicago: University of Chicago Press, 1996).

7. The illusion of mimesis, or imitating reality, has been challenged by many authors who argue that maps are not views but constructions. According to Christian Jacob: "A map is not a mimetic image, but an analogical image, the product of an abstraction that interprets the landscape and makes it intelligible. . . . A map thus constantly requires choices, exclusions, movements, and equivalence. . . . These operations are determined both by the codes of figuration and by the social and intellectual functions that motivate the drawing of the map." See Jacob, *The Sovereign Map*, 23.

8. The bird's-eye view and the horizontal view produce different interpretations of land. The bird's-eye view (*vue à vol d'oiseau*, or *Vogelschau*) has been called the "view from nowhere." Raymond Craib writes: "As the eye was detached from the viewer, surveying the landscape from above, so was it presumed that the map itself was disembodied, free of human bias and prejudice." See Raymond Craib, "Cartography and Power in the Conquest of New Spain," *Latin American Research Review* 35, no. 1 (Spring 2000): 20. The panoramic view, on the other hand, conveys a more experiential understanding of one's surroundings. Michel de Certeau, for example, contrasts the godlike perspective of an aerial-view map with that of a "tour," which is a "contextual and horizontal perspective derived from daily practices." See Michel de Certeau, *The Practice of Everyday Life*, trans. Steven E. Rendall (Berkeley: University of California Press, 1984), 118–21.

9. Beginning in the 1980s, Brian Harley and David Woodward founded a new school of "critical cartography" that revolutionized the study of historical maps. Eschewing the post-Enlightenment tendency to categorize maps in black-and-white terms as either scientifically accurate or wrong, Harley and Woodward sought to determine the meaning of maps in the context of the societies that made and used them. Throughout history, they argued, maps stored, communicated, and promoted spatial understandings according to the needs of each specific society. See J. B. Harley and David Woodward, eds., *The History of Cartography*, 6 vols. (Chicago: University of Chicago Press, 1987–present).

10. See, for example, Michael Biggs, "Putting the State on the Map: Cartography, Territory, and European State Formation," *Comparative Studies in Society and History* 41, no. 2 (April 1999): 374–405; Josef W. Konvitz, *Cartography in France, 1660–1848. Science, Engineering, and Statecraft* (Chicago: University of Chicago Press, 1987); David Buisseret, ed., *Monarchs, Ministers, and Maps: The Emergence of Cartography as a Tool of Government in Early Modern Europe* (Chicago: University of Chicago Press, 1992); Monique Pelletier, *Les cartes des Cassini: La science au service de l'État et des régions* (Paris: Éditions du C.T.H.S., 2002); James R. Akerman, "The Structuring of Political Territory in Early Printed Atlases," *Imago Mundi* 47 (1995): 138–54; James C. Scott, *Seeing like a State: How Certain Schemes to Improve the Human Condition Have Failed* (New Haven: Yale University Press, 1998).

11. In his chapter "Silences and Secrecy: The Hidden Agenda of Cartography in Early Modern Europe," Brian Harley argues that maps were intellectual weapons of the state system. See J. B. Harley, *The New Nature of Maps: Essays in the History of Cartography* (Baltimore: Johns Hopkins University Press, 2001), 88.

12. In his chapter "Maps, Knowledge, and Power," Brian Harley writes: "Maps as an impersonal type of knowledge tend to 'desocialize' the territory they represent. They foster the notion of a socially empty space. . . . Decisions about the exercise of power are removed from the realm of immediate face-to-face contacts." See Harley, *New Nature of Maps*, 81. Scholars of empire have adopted perhaps the strongest Foucauldian interpretation of maps. Historians of British imperialism, such as D. Graham Burnett and Matthew Edney, discuss British mapping efforts as a first step toward the imperial takeover of faraway lands. See Matthew H. Edney, *Mapping an Empire: The Geographical Construction of British India, 1765–1843* (Chicago: University of Chicago Press, 1997); D. Graham Burnett, *Masters of All They Surveyed: Exploration, Geography and a British El Dorado* (Chicago: University of Chi-

cago Press, 2000). The same perspective can be applied to empires within Europe. In her study on the Soviet Empire, Francine Hirsch calls maps a "cultural technology of rule" and an essential part of the "conceptual conquest of lands and peoples." See Francine Hirsch, *Empire of Nations: Ethnographic Knowledge and the Making of the Soviet Union* (Ithaca: Cornell University Press, 2005). Scholars of Europe's ancient empires have also used maps to better understand strategies of imperial control. See Claude Nicolet, *Space, Geography and Politics in the Early Roman Empire*, trans. Hélène Leclerc (Ann Arbor: University of Michigan Press, 1991). For an example of border-mapping practices in non-Western empires, see Peter Purdue, "Boundaries, Maps and Movement: The Chinese, Russian and Mongolian Empires in Early Modern Eurasia," *International History Review* 20, no. 2 (June 1998): 263–86.

13. In her analysis of Russian maps, Valerie Kivelson writes: "Uses and understandings of space were not just an important piece of a top-down enterprise of state conquest and control, but also a crucial area through which ordinary Muscovites made sense of, altered and negotiated for position in their world." See Valerie Kivelson, *Cartographies of Tsardom: The Land and Its Meanings in Seventeenth-Century Russia* (Ithaca: Cornell University Press, 2006), 8. For an insightful study on the intersection of cartography and Russian imperial identity during the eighteenth and nineteenth centuries, see Steven Seegel, *Mapping Europe's Borderlands: Russian Cartography in the Age of Empire* (Chicago: University of Chicago Press, 2012). Sumathi Ramaswamy, in her recent book, focuses on India's "barefoot cartographers"—untrained in the formalities of scientific surveying—rather than on the state's version of "command cartography." See Sumathi Ramaswamy, *The Goddess and the Nation: Mapping Mother India* (Durham: Duke University Press, 2010), 34–35. Scholars of American cartography have also explored the relationship between mapmaking and the formation of national identity. See Susan Schulten, *Mapping the Nation: History and Cartography in Nineteenth-Century America* (Chicago: University of Chicago Press, 2012). For an excellent discussion of map literacy in eighteenth-century America, see Martin Brückner, *The Geographic Revolution in Early America: Maps, Literacy, and National Identity* (Chapel Hill: University of North Carolina Press, 2006). For recent studies on the intersection of maps and identity in modern Europe, see Helmut Walser Smith's discussion of mapmaking and German identity formation in Helmut Walser Smith, *The Continuities of German History: Nation, Religion, and Race across the Long Nineteenth Century* (Cambridge: Cambridge University Press, 2008). See also Christof Dipper and Ute Schneider, eds., *Kartenwelten: Der Raum und seine Repräsentation in der Neuzeit* (Darmstadt: Primus, 2006). For a discussion of "topographies of the nation" in nineteenth-century Switzerland, see David Gugerli and Daniel Speich, *Topografien der Nation: Politik, kartografische Ordnung und Landschaft im 19. Jahrhundert* (Zurich: Chronos, 2002).

14. For scholarship addressing the popular dimensions of mapmaking, see Mark Heffernan, "The Cartography of the Fourth Estate: Mapping the New Imperialism in British and French Newspapers, 1875–1925," in *The Imperial Map: Cartography and the Mastery of Empire*, ed. James R. Akerman (Chicago: University of Chicago Press, 2009); Raymond B. Craib, *Cartographic Mexico: A History of State Fixations and Fugitive Landscapes* (Durham: Duke University Press, 2004); Marcia Yonemoto, *Mapping Early Modern Japan: Space, Place and Culture in the Tokugawa Period (1603–1868)* (Berkeley: University of California Press, 2003).

15. For a discussion of popular science in the nineteenth century, see Bernard Lightman, *Victorian Popularizers of Science: Designing Nature for New Audiences* (Chicago: University of Chicago Press, 2007).

16. I borrow the terms "high" and "low" cartography from Robert Darnton, who distinguished between the "high" Enlightenment of Voltaire and Rousseau and the "low" Enlightenment of Grub Street scribblers in eighteenth-century France. See Robert Darnton, *The Literary Underground of the Old Regime* (Cambridge, Mass.: Harvard University Press, 1985).

17. On the question of audience reception in visual history, see Michael L. Wilson, "Visual Culture: A Useful Category of Historical Analysis?" in *The Nineteenth-Century Visual Culture Reader*, ed. Vanessa Schwartz and Jeannene Przyblyski (New York: Routledge, 2004), 26–35. See also Peter Burke, *Eyewitnessing: The Uses of Images as Historical Evidence* (Ithaca: Cornell University Press, 2001).

18. On the history of print circulation, see Robert Darnton, *The Forbidden Best-Sellers of Pre-revolutionary France* (New York: W. W. Norton, 1996); Darnton, *Literary Underground*. For an excellent discussion of how map literacy works, see Brückner, *Geographic Revolution*, 1–15.

19. See Benjamin, "The Work of Art in the Age of Mechanical Reproduction"; for a discussion of the technologies that underpinned the methods of modern visual observation, see Jonathan Crary, *Techniques of the Observer: On Vision and Modernity in the Nineteenth Century* (Cambridge, Mass.: MIT Press, 1990).

20. On an expanded understanding of "mapness," see James R. Akerman and Robert W. Karrow Jr., eds. *Maps: Finding Our Place in the World* (Chicago: University of Chicago Press, 2007), 2.

21. Thongchai Winichakul defines the geo-body as "the operations of the technology of territoriality which created nationhood spatially." See Winichakul, *Siam Mapped*, 16.

22. See chap. 10, "Census, Map and Museum," of Benedict Anderson's influential study on nationalism, *Imagined Communities: Reflections on the Origin and Spread of Nationalism* (London: Verso, 1991), 163–85.

23. See, for example, Ramaswamy, *Goddess and the Nation*; Susan Schulten, *The Geographical Imagination in America, 1880–1950* (Chicago: University of Chicago Press, 2001).

24. For a classic example of the "modernization theory" approach to European nation building, see Eugen Weber, *Peasants into Frenchmen: The Modernization of Rural France, 1870–1914* (Stanford: Stanford University Press, 1976).

25. On the subject of nation building in provincial France and Germany, see Peter Sahlins, *Boundaries: The Making of France and Spain in the Pyrenees* (Berkeley: University of California Press, 1989); Caroline Ford, *Creating the Nation in Provincial France: Religion and Political Identity in Brittany* (Princeton: Princeton University Press, 1993); Timothy Baycroft, *Culture, Identity and Nationalism: French Flanders in the Nineteenth and Twentieth Centuries* (Suffolk: Royal Historical Society, Boydell Press, 2004); Patrick Young, *Enacting Brittany: Tourism and Culture in Provincial France, 1871–1939* (London: Ashgate, 2012); Jean-François Chanet, *L'École républicaine et les petites patries* (Paris: Aubier, 1996); Anne-Marie Thiesse, *Ils apprenaient la France: L'exaltation des régions dans le discours patriotique* (Paris: Éditions de la Maison des Sciences de l'Homme, 1997); Celia Applegate, *A Nation of Provincials: The German Idea of Heimat* (Berkeley: University of California Press, 1990); Alon Confino, *The Nation as a Local Metaphor: Württemberg, Imperial Germany, and National Memory, 1871–1918* (Chapel Hill: University of North Carolina Press, 1997).

26. In emphasizing the spatial foundations of nation building, I build upon the work of scholars who have argued that space is much more than the static "backdrop" to history, but is rather something as dynamic and changing as time. See Henri Lefebvre, *The Production of Space*, trans. D. Nicholson-Smith (Oxford: Basil Blackwell, 1991); Edward Soja, *Postmodern Geographies: The Reassertion of Space in Critical Social Theory* (London: Verso, 1989); Erica Carter et al., eds., *Space and Place: Theories of Identity and Location* (London: Lawrence and Wishart, 1993); Doreen Massey, *For Space* (London: Sage Press, 2005); Alan Baker, *Geography and History: Bridging the Divide* (Cambridge: Cambridge University Press, 2003); Robert David Sack, *Human Territoriality: Its Theory and History* (Cambridge: Cambridge University Press, 1986). For recent examples of spatial histories focusing on modern Germany, see Karl Schlögel, *Im Raume Lesen Wir die Zeit* (Munich: Carl Hanser Verlag, 2003); Thomas Kühne, ed., *Raum und Geschichte* (Leinfelden-Echterdingen: DRW-Verlag, 2001); David Blackbourn, *The Conquest of Nature: Water, Landscape, and the Making of Modern Germany*

(Cambridge, Mass.: Harvard University Press, 2006); Dipper and Schneider, *Kartenwelten*. For modern France, see selections from Pierre Nora, ed., *Rethinking France: Les Lieux de Mémoire*, trans. Mary Seidman Trouille, 3 vols. (Chicago: University of Chicago Press, 2001–6); Fernand Braudel, *La Méditerranée et le monde méditerranéen à l'époque de Philippe II* (Paris: Colin, 1949); Fernand Braudel, *L'identité de la France: Espace et territoire* (Paris: Flammarion, 1986); Jacques Revel, ed., *L'espace français* (Paris: Editions du Seuil, 1989); Anne Godlewska, *Geography Unbound: French Geographic Science from Cassini to Humboldt* (Chicago: University of Chicago Press, 1994).

27. Like Rogers Brubaker and Frederick Cooper, I find the term "national identification" to be helpful in describing the process of becoming national. For their incisive article on the subject, see Rogers Brubaker and Frederick Cooper, "Beyond 'Identity,'" *Theory and Society* 24, no. 1 (February 2000): 1–47.

28. See Kären Wigen, *A Malleable Map: Geographies of Restoration in Central Japan, 1600–1912* (Berkeley: University of California Press, 2010).

29. On the revolutionary remapping of France, see Marie-Vic Ozouf-Marignier, *La formation des départements: La représentation du territoire française à la fin du 18e siècle* (Paris: Éditions de l'École des hautes études en sciences sociales, 1989).

30. On the resurrection of local and regional space in nineteenth-century France, see Stéphane Gerson, *The Pride of Place: Local Memories and Political Culture in Nineteenth-Century France* (Ithaca: Cornell University Press, 2003); Philip Whalen and Patrick Young, eds., *Place and Locality in Modern France* (New York: Bloomsbury Press, 2014).

31. On the regional movement in Alsace, see Christopher Fischer, *Alsace to the Alsatians? Visions and Divisions of Alsatian Regionalism, 1870–1939* (New York: Berghahn, 2010); Bernard Vogler, *Histoire culturelle de l'Alsace: Du Moyen Age à nos jours, les très riches heures d'une région frontière* (Strasbourg: La Nuée bleue, 1994); François Igersheim, *L'Alsace et ses historiens, 1680–1914: La fabrique des monuments* (Strasbourg: Presses universitaires de Strasbourg, 2006). On regional identity in Lorraine, see David Hopkin, "Identity in a Divided Province: The Folklorists of Lorraine, 1860–1960," *French Historical Studies* 23, no. 4 (2000): 639–82.

32. On Alsace-Lorraine's shifting legal status as a state within the German Empire, see Dan P. Silverman, *Reluctant Union: Alsace-Lorraine and Imperial Germany, 1871–1918* (University Park: Pennsylvania State University Press, 1972); Hans-Ulrich Wehler, "Unfähig zur Verfassungsreform: Das 'Reichsland' Elsass-Lothringen von 1870 bis 1918," in *Krisenherde des Kaiserreichs 1871–1918* (Göttingen: Vandenhoeck und Ruprecht, 1970), 17–64; Alfred Wahl and Jean-Claude Richez, *L'Alsace entre France et Allemagne, 1850–1950* (Paris: Hachette, 1993). For a discussion of German and French educational policy in disputed Alsace-Lorraine, see Stephen Harp, *Learning to Be Loyal: Primary Schooling as Nation Building in Alsace and Lorraine, 1850–1940* (DeKalb: Northern Illinois University Press, 1998); John E. Craig, *Scholarship and Nation Building: The Universities of Strasbourg and Alsatian Society, 1870–1939* (Chicago: University of Chicago Press, 1984). On issues of religion, class, nationality, and gender in disputed Alsace-Lorraine, see Vicki Caron, *Between France and Germany: The Jews of Alsace-Lorraine, 1871–1918* (Stanford: Stanford University Press, 1988); David Allen Harvey, *Constructing Class and Nationality in Alsace, 1830–1945* (DeKalb: Northern Illinois University Press, 2001); Anthony Steinhoff, *The Gods of the City: Protestantism and Religious Culture in Strasbourg, 1870–1914* (Boston: Brill, 2008); Elizabeth Vlossak, *Marianne or Germania? Nationalizing Women in Alsace, 1870–1946* (Oxford: Oxford University Press, 2010).

33. On the concept of "counter-mapping," see Peluso, "Whose Woods Are These?"

34. For a detailed study of Alsace-Lorraine's language demographics, see Paul Lévy, *Histoire linguistique de l'Alsace et de Lorraine* (Strasbourg: Imprimerie alsacienne, 1929), vol. 2.

35. For comparative studies on modern France and Germany, see Rogers Brubaker, *Citi-*

zenship and Nationhood in France and Germany (Cambridge, Mass.: Harvard University Press, 1992); Dieter Gosewinkel, *Figurationen des Staates in Deutschland und Frankreich, 1870–1945* (Munich: R. Oldenbourg, 2006); Hartmut Kaelble, *Nachbarn am Rhein: Entfremdung und Annäherung der französischen und deutschen Gesellschaft seit 1880* (Munich: Beck, 1991).

36. On the emerging field of transnational history, see C. A. Bayly et al., "AHR Conversation: On Transnational History," *American Historical Review* 111, no. 5 (December 2006): 1441–64; Gunilla Budde et al., eds., *Transnationale Geschichte: Themen, Tendenzen und Theoren* (Göttingen: Vandenhoeck und Ruprecht, 2006); Akira Iriye, "Transnational History," *Contemporary European History* 13, no.2 (2004): 211–22.

Chapter 1

1. Bruno Latour argues that a person's ability to visualize absent things is a potent form of modern power. "A great man," he writes, "is a little man looking at a good map." See Bruno Latour, "Visualization and Cognition: Drawing Things Together," *Knowledge and Society: Studies in the Sociology of Culture Past and Present* 6 (1986): 26.

2. For studies on the relationship between cartography and the making of modern European states, see Michael Biggs, "Putting the State on the Map: Cartography, Territory, and European State Formation," *Comparative Studies in Society and History* 41, no. 2 (April 1999): 374–405; Joseph W. Konvitz, *Cartography in France, 1660–1848: Science, Engineering, and Statecraft* (Chicago: University of Chicago Press, 1987); Monique Pelletier, *Les cartes des Cassini: La science au service de l'État et des regions* (Paris: Éditions du C.T.H.S., 2002); James R. Akerman, "The Structuring of Political Territory in Early Printed Atlases," *Imago Mundi* 47 (1995): 138–54. For a political scientist's perspective, see James C. Scott, *Seeing like a State: How Certain Schemes to Improve the Human Condition Have Failed* (New Haven: Yale University Press, 1998).

3. See James Duncan and David Ley, eds., *Place/Culture/Representation* (London: Routledge, 1993), 2. See also Michel Foucault, *Discipline and Punish* (New York: Vintage, 1995), 195.

4. See introduction, n. 12.

5. A cross-national comparison of Alsace-Lorraine's maps does, however, create a diachronic problem: the French state undertook its first comprehensive map survey in the eighteenth century, while the Germans did not create their first unified imperial map until the late nineteenth century. The historian must therefore be attuned to the different timelines for state formation in the two countries. On the issue of diachronic comparison, see Heinz-Gerhard Haupt and Jürgen Kocka, "Comparative History: Methods, Aims, Problems," in *Comparison and History: Europe in Cross-National Perspective*, ed. Maura O'Connor and Deborah Cohen (New York: Routledge, 2004); Jürgen Kocka, "Comparison and Beyond," *History and Theory* 42 (February 2003): 39–44. See also Marc Bloch's views on time and comparison in a reprinting of his foundational 1928 discussion of comparative historical methodology: Marc Bloch, "Toward a Comparative History of European Societies," in *Enterprise and Secular Change: Readings in Economic History*, ed. Frederic C. Lane and Jelle C. Riemersma (Homewood, Ill.: R. D. Irwin, 1953).

6. For other examples of transnational cooperation in border mapping, see Paula Rebert, *La Gran Linea: Mapping the United States-Mexico Boundary, 1849–1857* (Austin: University of Texas Press, 2001); Rachel St. John, *Line in the Sand: A History of the Western U.S.-Mexico Border* (Princeton: Princeton University Press, 2012).

7. On the history of cross-border transfer between modern France and Germany, see Michel Espagne, *Les transfers culturels franco-allemands* (Paris: Presses universitaires de France, 1999), 269. On the concept of Europe's "entangled history," see Michel Werner and Bénédicte Zimmermann, "Beyond Comparison: *Histoire croisée* and the Challenge of Reflexivity," *History and Theory* 45, no. 1 (February 2006): 30–50.

8. See Scott, *Seeing like a State*, 83.

9. The Bibliothèque nationale et universitaire de Strasbourg holds a German version of Specklin's map of Upper Alsace and a Latin version of Specklin's map of Lower Alsace. See Daniel Specklin, *Alsatia Superior (Ober Elsas)*, 1592, Bibliothèque nationale et universitaire de Strasbourg, M.Carte.100.108; Daniel Specklin, *Alsatia Inferior (Under Elsas)*, 1592, Bibliothèque nationale et universitaire de Strasbourg, M.Carte.100.011.

10. See Albert Fischer, *Daniel Specklin aus Strassburg: Festungsbaumesiter, Ingenieur und Kartograph* (Sigmarien: Jan Thorbecke Verlag, 1996), 179.

11. François de Dainville, "L'Alsace comme la voyaient les cartes anciennes," *Saisons d'Alsace* 22 (1967): 153–76.

12. See the text at the bottom of Specklin, *Alsatia Inferior*, 1592.

13. Ch. Schott, "Daniel Specklin und seine Karte des Elsasses," *A Travers les Vosges* 16 (1920).

14. See Peter Sahlins, "Natural Frontiers Revisited: France's Boundaries since the Seventeenth Century," *American Historical Review* 95, no. 5 (December 1990): 1423–51.

15. Konvitz, *Cartography in France*, 150. See also Pelletier, *Les cartes des Cassini*, 35.

16. As numerous historians have recently argued, Old Regime France exhibited both stagnating and modernizing impulses. Cartography represented one of the most significant innovating drives of the Old Regime. See, for example, Christine Marie Petto, *When France Was King of Cartography: The Patronage and Production of Maps in Early Modern France* (New York: Lexington Books, 2007).

17. See Cassini III, *Nouvelle carte qui comprend les principaux triangles qui servent de fondement à la description géométrique de la France*, 1744, Bibliothèque nationale de France, Ge BB 565 A (VII), plate 10.

18. See Cassini III's explanation of the scientific methods that he used to conduct his map survey. Cassini de Thury, *Avertissement ou introduction à la carte générale et particulière de la France*, 1744, Bibliothèque nationale de France, Ge FF 4296.

19. Ibid.

20. Explaining the modern attitude toward mapmaking first established by the Cassinis, General Berthaut, the head of the Geographical Service of the French Army, wrote: "[Cartography] is not about a more or less mysterious gift of guessing or intuition, but rather the application of a science, as well as the constant practice of reasoned observation. . . . To know the terrain, you must necessarily understand it, and to understand it, you must reason it." See Henri Berthaut, *Connaissance du terrain et lecture des cartes* (Paris: Imprimerie du Service géographique de l'armée, 1912), 75.

21. See David Gugerli and Daniel Speich, *Topografien der Nation: Politik, kartographische Ordnung und Landschaft im 19. Jahrhundert* (Zurich: Chronos, 2002), 11.

22. See Cassini, *Avertissement*, 9. For an account of an incident between a surveyor and locals in Alsace, see Konvitz, *Cartography in France*, 15. See also Graham Robb's chapter "The Undiscovered Continent," in *The Discovery of France: A Historical Geography from the Revolution to the First World War* (New York: W. W. Norton, 2007), 3–18.

23. See Peter Sahlins, *Boundaries: The Making of France and Spain in the Pyrenees* (Berkeley: University of California Press, 1989), 94–96.

24. See D'Arçon, *Carte des frontières de l'Est*, Bibliothèque nationale de France, Ge DD 5464 (17). For a discussion of d'Arçon's border mapping project, see Konvitz, *Cartography in France*, 33.

25. "Immutability," argues Latour, "is ensured by the process of printing many identical copies; mobility by the number of copies, the paper and the movable type." See Latour, "Visualization and Cognition," 10.

26. Cassini, *Avertissement*, 4.

27. See Henri Berthaut, *Les ingénieurs-géographes militaires, 1624–1831*, vol. 1 (Paris: Imprimerie du Service géographique de l'armée, 1902), 142.

28. Napoleon's Topographical Commission abandoned the former royal measurement

system in favor of the metric system with decimal scales and adopted a new system for representing topographical relief that included hash marks and curves for defining slopes. See Pelletier, *Les cartes des Cassini*, 30. See also Monique Pelletier and Henriette Ozanne, *Portraits de la France: Les cartes, témoins de l'histoire* (Paris: Hachette, 1995).

29. A similar form of French-German cooperation took place in Bavaria. The French surveyor Bonne performed triangulation work and found astronomical points, while the German von Riedel worked with Bavarian geometers to complete Bavaria's Napoleonic map. Bonne wrote to the Dépôt de la Guerre in Paris, commending the Bavarian engineers for their great skill with the measuring table. See Rudiger Finsterwalder, *Zur Entwicklung der bayerischen Kartographie von ihren Anfängen bis zum Beginn der amtlichen Landesaufnahme* (Munich: Verlag der Bayerischen Akademie der Wissenschaften, 1967).

30. See Georg Krauss, "150 Jahre preussische Messtischblätter," *Zeitschrift für Vermessungswesen* 4 (April 1969): 125–35.

31. See Alphons Habermeyer, *Die topographische Landesaufnahme von Bayern im Wandel der Zeit* (Stuttgart: Verlag Konrad Wittwer, 1993), 114.

32. *Topographische Karte des Rheinstromes und seiner beiderseitigen Ufer von Hueningen bis Lauterburg oder längs der Französische-Badischen Gränze* (Freiburg im Breisgau: Herderschen Kunst Institut, 1828), in Bibliothèque nationale de France, Ge CC 27 42.

33. See written commentary, ibid.

34. For a recent discussion of the importance of canals to German identity, see David Blackbourn, *The Conquest of Nature: Water, Landscape, and the Making of Modern Germany* (Cambridge, Mass.: Harvard University Press, 2006).

35. See letter from the commanding French general Jean-Jacques Alexis Uhrich to Prefect Plon, 18 August 1870, Archives départementales du Bas-Rhin, 14 M 2. I thank Rachel Chrastil for drawing my attention to this letter.

36. The Department of the Moselle comprised the arrondissements of Metz, Thionville, Sarreguemines, Château-Salins, and Sarrebourg.

37. *Carte de la frontière franco-allemande: 1870–1871 (1:80,000)*, Archives départementales du Bas-Rhin, 1J/Plan 1.

38. See Georges Delahache, *Alsace-Lorraine: La carte au liseré vert* (Paris: Hachette, 1918).

39. Quoted in ibid., 196–97.

40. *Notizen über die Thätigkeit der internationalen Grenz-Regulierungs-Commission und den Verlauf der Grenz-Arbeiten*, August 1872, Archives départementales du Bas-Rhin, 1J/Plan1.

41. Archives départementales du Bas-Rhin, 71 AL 7.

42. *Frontière franco-allemande: Plan d'ensemble; Établie par les géometres en chef Laloy pour la France et Husangen pour l'Allemagne*, 1871, Archives départementales de Meurthe-et-Moselle, 1 Fi 21. During the nineteenth century, this kind of joint border-mapping procedure became widespread among states across the globe. In her book, Paula Rebert demonstrates that the heads of the US and Mexican boundary commissions mostly agreed on cartographic standards and approved each other's boundary maps with official signatures. See Rebert, *La Gran Linea*, 54–55.

43. *Procès-verbal de délimitation entre la France et l'Allemagne*, 26 April 1877, Archives départementales du Bas-Rhin, 87 AL 1142.

44. See Archives départementales de Meurthe-et-Moselle, 4 M 175; and Archives départementales du Bas-Rhin, AL 64.

45. See letters of protest in Archives départementales du Bas-Rhin, 1 AL 64.

46. For a detailed discussion of the French-German border administration from 1871 to 1918, see François Roth, "La frontière franco-allemande 1871–1918," in *Grenzen und Grenzregionen*, ed. Wolfgang Haubrichs and Reinhard Schneider (Saarbrücken: Saarbrücker Druck und Verlag, 1993), 131–45.

47. For reports on the upkeep of border stones, see Archives départementales de Meurthe-et-Moselle, 4 M 176.

48. In his study on the Anglo-Gorkha (now India-Nepal) border, Bernardo Michael documents the challenges of establishing a linear boundary across a territory with interconnected local communities. See Bernardo Michael, *Statemaking and Territory in South Asia: Lessons from the Anglo-Gorkha War (1814–1816)* (London: Anthem Press, 2012), 85.

49. O. von Morozowicz, *Die Königlich Preussische Landes-Aufnahme* (Berlin: Königlich Preussische Landes-Aufnahme, 1878), 1.

50. During the 1830s and 1840s, cartographers from Baden, Bavaria, Württemberg and Hesse had attempted to create a similar comprehensive map that unified all of their state territories on paper. They could not agree on measurement techniques, and the project failed.

51. For an example, see Königlich Preussischen Landesaufnahme, *Musterblatt und Zeichenerklärung für die topographischen und kartographischen Arbeiten im Masstabe 1:25,000* (Berlin: Königlich Preussische Landesaufname, 1913).

52. Reichskanzleramt to Strassburg, 20 March 1872, Archives départementales du Bas-Rhin, 87 AL 1142.

53. J. Schroeder-Hohenwarth, "Die preussische Landesaufnahme von 1816–1875," *Nachrichten aus dem Karten- und Vermessungswesen* 5 (1958): 37.

54. Von Moltke to the German administration in Strassburg, 9 March 1872, Archives départementales du Bas-Rhin, 87 AL 1142.

55. *Erinnerungsblatt des Kaisermanövers des XV. Armee Corps 1886*, Bibliothèque nationale et universitaire de Strasbourg, M.Carte.10202.

56. Königlich Preussischen Landesaufnahme, *Die Arbeiten der Königlich Preussischen Landes-Aufnahme* (Berlin: Königlich Preussischen Landesaufnahme, 1893), 13.

57. "Neue topographische Spezialkarte von Elsass-Lothringen im Massstabe 1:80,000," *Militär Wochenblatt*, no. 103 (December 1878).

58. See copies of requests in Archives départementales du Bas-Rhin, 87 AL 3374.

59. *Karte von Elsass-Lothringen auf Grund besonderer Recognosierungen unter Benutzung der "Carte de France (1:80,000),"* Bibliothèque municipale de Strasbourg, A289.

60. See map of *Kreis Metz* in the bound volume entitled *Uebersichtskarten der Triangulation 3. & 4. Ordnung (1:100,000)*, Archives départementales du Bas-Rhin, 4K10. For a discussion of how the Germans used French trigonometric points, see the letter to the Finance Ministry of Alsace-Lorraine, 4 September 1884, Archives départementales du Bas-Rhin, 87 AL 1142.

61. See Archives départementales du Bas-Rhin, 87 AL 3374.

62. There is evidence that such confiscations of maps also took place during the Franco-Prussian War of 1870–71, albeit on a smaller scale. Commanding General Jean-Jacques Alexis Uhrich wrote a letter to the publishing house Berger-Levrault in August 1870, asking the company to suspend publication of its plans of Strasbourg while the Germans held the city under siege. The letter also expressed General Uhrich's fears that "the vast majority of these plans have left the city to be used for hostile purposes." Uhrich himself was said to have borrowed his battle map from a local lawyer. Letter from Uhrich to Prefect Pron, 18 August 1870, Archives départementales du Bas-Rhin, 14 M 2. I thank Rachel Chrastil for drawing my attention to this reference.

63. For a list of confiscated maps, see *Verzeichnis derjenigen Kartenwerke, deren Betrieb im Mobilmachungsfälle im militarischen Interesse einzustellen ist*, July 1910, Archives départementales du Bas-Rhin, 30 AL 54. Commercial maps did play a strategic role in the war. French intelligence reported in 1915 that German troops were using bicycling maps that they found in occupied villages in France. See Captain Renaud to the Service géographique de l'armée, 10 January 1915, Service historique de la défense, Vincennes, 3M577. In February 1915, the French War Ministry requested the National Library in Paris to locate the Vogesenclub's maps of the Alsatian village of Thann. See Service historique de la défense, Vincennes, 3M576.

64. Written correspondence between the Verein Elsass-Lothringischer Buchhandler and the government of Alsace-Lorraine continued throughout 1912. See Archives départementales du Bas-Rhin, 87 AL 5066.

65. See the minutes of the Service géographique de l'armée, 7 January 1915, Service historique de la défense, Vincennes, 3M577. French restrictions on the sale of maps had also reversed an earlier trend toward the public sale of maps. In 1870, the Dépôt de la Guerre, the military bureau in charge of printing most of the maps in France for both state and public uses, produced a total of 24,000 maps. In 1873, that figure rose to 92,000; in 1879 it rose to 263,000; by 1910 it had risen to 956,615 maps (598,793 of which were sold through private bookstores and distributors). Statistics cited in Service géographique de l'armée, *Rapports sur les travaux exécutés* (Paris: Librairie Militaire de L. Baucoin, 1910), 79. For a description of the map restrictions for the entire German territory during World War I, see *Verordnung über Ausfuhr und Vertrieb von Karten und Geländebeschreibungen*, 9 April 1918, Service historique de la défense, Vincennes, 3M576.

66. For a discussion of militarized visions of land on the eastern front, see Vejas Liulevicius, *War Land on the Eastern Front: Culture, National Identity, and German Occupation in World War I* (Cambridge: Cambridge University Press, 2000).

67. See Ministère de la Guerre, *Instruction sur les cartes et plans directeurs de guerre (première partie)* (Paris: Service géographique de l'armée, 1925), 14.

68. Intelligence gathered from interrogations of German prisoners indicated that military personnel lower than the rank of captain were not granted access to maps. See Service historique de la défense, Vincennes, 3M577.

69. German army deserters from Lorraine reported to French military intelligence that battlefield plans on the scale of 1:20,000 or 1:10,000 found on dead bodies or prisoners sold on the black market for ten to twenty marks. Head of the French Second Army's Groupe de canevas de tir to the director of the Service géographique de l'armée, 10 January 1917, Service historique de la défense, Vincennes, 3M576.

70. *Note sur des cartes et plans speciaux à exécuter dans les armées*, 10 August 1917, Service historique de la défense, Vincennes, 3M583.

71. Michel Espagne notes that the seizure or transfer of documents and archives across borders was an important point of cultural transfer between modern France and Germany. See Espagne, *Les transfers culturels franco-allemands*, 86.

72. *Carte secrète de l'état-major de la région de Sainte-Marie-aux-Mines*, 1 September 1918, Archives départementales du Bas-Rhin, 7K3. For a map of the Hartmannswillerkopf area during an active period of battle in March 1915, see Service historique de la défense, Vincennes, 3M577. For a written description of the French army's use of German maps, see Lieutenant-Colonel Perrier, *Historique du Groupe de Canevas de Tir de la VII Armée*, 1923, Service historique de la défense, Vincennes, 3M561.

73. *Situation économique en pays ennemis: La pénurie de papier en Allemagne*, 6 February 1918, Service historique de la défense, 3M576.

74. Ministère de la Défense Nationale et de la Guerre, *Le Service géographique de l'armée: Son histoire, son organisation, ses travaux* (Paris: Imprimerie du Service géographique de l'armée, 1938).

75. For a discussion of the destructive twentieth-century consequences of Enlightenment thinking, see Reinhart Koselleck, *Kritik und Krise* (Freiburg: K. Alber, 1959).

76. Guides Illustrés Michelin des Champs de Bataille, *Metz et la bataille de Morhange* (Clermont-Ferrand: Michelin & Co., 1919), 40.

77. Guides Illustrés Michelin des Champs de Bataille, *L'Alsace et les combats des Vosges, 1914–1918* (Clermont-Ferrand: Michelin & Co., 1919), 72.

78. Ibid., 69–71.

79. Instructions on making provisional maps for Alsace-Lorraine, 18 March 1919, Archives départementales du Bas-Rhin, 121 AL 738.

80. *Convention avec Monsieur le Commandant Paul Poliacchi, pour l'établissement de la Carte de l'Alsace et la Lorraine à l'échelle du 1:100,000*, Archives départementales du Bas-Rhin, 121 AL 738.

81. Instructions from the minister of war to the commissioner of the republic in Strasbourg, 23 May 1922, Archives départementales du Bas-Rhin, 121 AL 738.

82. Revisions for maps of the Spanish border had been projected for 1940–43. See Service historique de la défense, Vincennes, 9N296.

83. *Instruction relative aux cartes de couverture destinées aux États-Majors et troupes des régions fortifiées, secteurs fortifiés et defensifs, ainsi qu'aux troupes des garnisons de sûreté de certaines villes et positions de barrage*, 18 June 1934, Service historique de la défense, Vincennes, 9N296.

84. See report from the Army Geographical Service's syndicate of cartographers, July 1932, Service historique de la défense, Vincennes, 6N500.

Chapter 2

1. In analyzing how French and German publics mapped their national borders, this study builds on the work of recent historians of nationalism who have stressed the importance of civil society to the construction of national identities. See, for example, Kristen Belgum, *Popularizing the Nation: Audience, Representation, and the Production of Identity in "Die Gartenlaube," 1853–1900* (Lincoln: University of Nebraska Press, 1998); Pieter Judson, *Guardians of the Nation: Activists on the Language Frontiers of Imperial Austria* (Cambridge, Mass.: Harvard University Press, 2006); Thomas Serrier, *Entre Allemagne et Pologne: Nations et identités frontalières, 1848–1914* (Paris: Belin, 2002).

2. For "geo-body," see Thongchai Winichakul, *Siam Mapped: A History of the Geo-body of a Nation* (Honolulu: University of Hawaii Press, 1994); for "mythical space," see Yi-Fu Tuan, *Space and Place: The Perspective of Experience* (Minneapolis, University of Minnesota Press, 2001), 99; for "imaginative geography," see Edward W. Said, *Orientalism* (New York, Vintage, 1979).

3. See chap. 10, 'Census, Map and Museum,' in Benedict Anderson, *Imagined Communities: Reflections on the Origin and Spread of Nationalism* (London: Verso, 1991), 163–85.

4. Helmut Walser Smith has argued that Germans used their senses to "discover" their nation in the land, rather than imagining their nation in more abstract terms. See Helmut Walser Smith, *The Continuities of German History: Nation, Religion, and Race across the Long Nineteenth Century* (Cambridge: Cambridge University Press, 2008), 73.

5. For an introduction to the concept of "thematic mapping" and its links to nineteenth-century social science, see Susan Schulten, *Mapping the Nation: History and Cartography in Nineteenth-Century America* (Chicago: University of Chicago Press, 2012), 1–4. See also Gilles Palsky, *Des chiffres et des cartes: La cartographie quantitative au XIXe siècle* (Paris: Comité des travaux historiques et scientifiques, 1996).

6. For an excellent discussion of the links between nationalism and the public research university in modern Europe, see John E. Craig, *Scholarship and Nation Building: The Universities of Strasbourg and Alsatian Society, 1870–1939* (Chicago: University of Chicago Press, 1984).

7. The nineteenth-century commercial map market built upon an earlier trend toward the public circulation of cartographic images in eighteenth-century Europe. See Mary Sponberg Pedley, *The Commerce of Cartography: Making and Marketing Maps in Eighteenth-Century France and England* (Chicago: University of Chicago Press, 2005).

8. The term "public sphere," popularized through the writings of Jürgen Habermas, has shaped recent discussions of eighteenth-century French social and cultural history. The term refers to an exclusive, educated, male public that engaged in "enlightened" discourses with the goal of reforming the French state. See Jürgen Habermas, *The Structural Transformation of the Public Sphere*, trans. Thomas Burger (Cambridge, Mass.: MIT Press, 1989).

9. Jean-Jacques Rousseau, quoted in Peter Sahlins, "Natural Frontiers Revisited: France's Boundaries since the Seventeenth Century," *American Historical Review* 95, no. 5 (December 1990): 1436.

10. The Cassinis' *Map of France* (1744) emphasized mountain and river iconography. For a detailed analysis of the "natural borders" concept in European territories overseas, see Charles Cheney Hyde, "Notes on Rivers as Boundaries," *American Journal of International Law* 6 (1912): 901–9.

11. Sahlins, "Natural Frontiers Revisited," 1447.

12. Ibid., 1440.

13. Danton's speech to the convention from 31 January 1793. Quoted in Jean-Marie Mayeur, "A Frontier Memory: Alsace," in *Rethinking France: Les Lieux de Mémoire*, ed. Pierre Nora, trans. Mary Seidman Trouille, 3 vols. (Chicago: University of Chicago Press, 2006), 2:418.

14. Image from the cover of Georges-Guillaume Boehmer, *La rive gauche du Rhin, limite de la République française, ou Recueil de plusieurs dissertations dignes des prix proposés par un négociant de la rive gauche du Rhin* (Paris: Chez Desenne, Louvet et Devaux, year IV). A copy of this rare book can be found at the Bibliothèque municipale de Strasbourg. This pamphlet is mentioned in Daniel Nordman, *Frontières de la France: De l'espace au territoire, XVI–XIXe siècle* (Paris: Gallimard, 1998), 128. See also Sahlins, "Natural Frontiers Revisited," 1447.

15. See Boehmer, *La rive gauche du Rhin*, 1.

16. Ibid., 51.

17. See Jacob Grimm and Wilhelm Grimm, *Märchen der Brüder Grimm* (Weinheim: Beltz und Gelberg, 1995).

18. Heinrich Berghaus, *Physikalischer Atlas*, 2 vols. (Gotha: Justus Perthes, 1848), vol. 2.

19. The French Geographical Society, founded in 1821, was the oldest in Europe. Many of its founding members had accompanied Napoleon on his expeditions to Egypt. The German Geographical Society of Berlin, created by Heinrich Berghaus and Karl Ritter, was founded second, in 1828. The Royal Geographical Society of Great Britain followed in 1830, and the Russian Geographical Society followed in 1845. Outside Europe, powerful geographic societies formed in several Latin American countries, including Brazil and Mexico.

20. Referring to the development of the atlas, Christian Jacob writes: "Its advent attests to the new demand of a learned and wealthy public for the type of nonutilitarian volume needed to complete a geographical education and to familiarize its users during relaxed or studious consultation with geographical forms and names. It is also needed, perhaps, as a sign of culture and social 'distinction.'" See Christian Jacob, *The Sovereign Map: Theoretical Approaches in Cartography throughout History*, trans. Tom Conley (Chicago: University of Chicago Press, 2006), 67.

21. Jürgen Espenhorst, *Petermann's Planet: A Guide to German Handatlases and Their Siblings throughout the World, 1800–1950*, vol. 1, trans. George R. Crossman (Schwerte: Pangaea Verlag, 2003), 67–68.

22. For a discussion of map semiotics and German nationalism, see Morgane Labbé, "Les frontières de la nation allemande dans l'espace de la carte, du tableau statistique et de la narration," in *L'espace de l'Allemagne au XIXe siècle: Frontières, centres et question nationale*, ed. Catherine Maurer (Strasbourg: Presses universitaires de Strasbourg, 2007).

23. See title page of Berghaus, *Physikalischer Atlas*, vol. 2.

24. Heinrich Berghaus, *Übersicht von Eüropa mit ethnograph. Begränzung der einzelnen Staaten, und den Völker-Sitzen in der Mitte des 19ten Jahrhunderts*, 1846, in Berghaus, *Physikalischer Atlas*, vol. 2.

25. For a comparative global example of ethnographic mapmaking and the classification of peoples, see Laura Hostetler, *Qing Colonial Enterprise: Ethnography and Cartography in Early Modern China* (Chicago: University of Chicago Press, 2001).

26. For a discussion of the semiotics of map subspaces, see Jacob, *The Sovereign Map*, 189.

27. For an analysis of nineteenth-century definitions of ethnicity in Germany, see H. Glenn Penny, *Objects of Culture: Ethnology and Ethnographic Museums in Imperial Germany* (Chapel Hill: University of North Carolina Press, 2002).

28. Berghaus notes that he used evidence from ancient monuments, as well as etymological research, to delineate ethnic borders in the British Isles. Berghaus, *Physikalischer Atlas*, vol. 2, no. 9, 16.

29. According to the footnotes in Berghaus's atlas, Heinrich Nabert's map of the French-German language border, *Über Sprachgrenzen insonderheit die deutsch-französischen in den Jahren 1844–1847*, served as an important basis for Berghaus's ethnic border between France and Germany. Ibid., 16.

30. According to Berghaus's notes, Nabert mapped the border between Gaelic and English by traveling from town to town, village to village, and house to house. Ibid., 17.

31. For a discussion of the "ethnoschematization" techniques that nineteenth-century Europeans used to invent national boundaries, see Seegel, *Mapping Europe's Borderlands*, 133–157.

32. Gilles Palsky is an expert on the graphic design of nineteenth-century statistical and thematic maps. For his discussion of ethnographic maps, see Gilles Palsky, *Des chiffres et des cartes: La cartographie quantitative au XIXe siècle* (Paris: Comité des travaux historiques et scientifiques, 1996), 251.

33. See Alain Blum, "Resistance to Identity Categorization in France," in *Census and Identity: The Politics of Race, Ethnicity and Language in National Censuses*, ed. Dominique Arel and David I. Kertzer (Cambridge: Cambridge University Press, 2002), 127; Dominique Arel, "Language Categories in Censuses: Backward-or-Forward-Looking?" in Arel and Kertzer, *Census and Identity*, 94.

34. For a discussion of Legoyt's statistical methodology, see Alfred Legoyt, *La France et l'étranger: Études d'une statistique comparée* (Strasbourg: Berger-Levrault,1865).

35. See Charles Schoebel, *La question d'Alsace au point de vue éthnographique* (Paris: Sandoz et Fischbacher, Éditeurs, 1872), 6–7.

36. Based on the map's handwritten notations, the image probably dates from July 1872 and was printed by the publishing house Franz Duncker in Berlin. *Die Deutsche Grenze gegen Frankreich: 25 Gedenkzeilen von Elsass und Lothringen*, Bibliothèque nationale et universitaire de Strasbourg, M.Carte.259.

37. Paul Claval argues that geography's increased popularity in late nineteenth-century French society was a reaction to France's disasterous military defeat in 1871. See Paul Claval, *Histoire de la géographie française de 1870 à nos jours* (Paris: Nathan, 1998), 19.

38. On the historical evolution of the hexagon, see Nathaniel B. Smith, "The Idea of the French Hexagon," *French Historical Studies* 6, no. 2 (Autumn 1969): 139–55. See also Eugen Weber's chapter "In Search of the Hexagon," in *My France: Politics, Culture, Myth* (Cambridge, Mass.: Harvard University Press, 1991), 57–71.

39. F. Schrader and L. Galloúdec, *Géographie de la France et de ses colonies* (Paris: Hachette, 1899), 4–5.

40. While historian Todd Shepard rightly argues that the "hexagon" served as a useful visual tool for distinguishing postcolonial France from its lost empire following the Algerian War, it is important to recognize that the hexagon played a different role in French political culture during the late nineteenth century and that its rise as a popular visual icon was deeply connected to the loss of Alsace-Lorraine after 1871. See Todd Shepard, *The Invention of Decolonization: The Algerian War and the Remaking of France* (Ithaca: Cornell University Press, 2006), 101.

41. See description of map 9 in F. Schrader, *Atlas de géographie moderne* (Paris: Hachette, 1890).

42. E. Levasseur and Ch. Périgot, *Cartes pour server à l'intelligence de la France avec ses colonies* (Paris: Ch. Delagrave, 1874).

43. For a collection of photographs from the 1875 geographic exposition in Paris, see Bibliothèque nationale de France, SG W2.

44. Société de Géographie, *Congrès international des sciences géographiques tenu à Paris du 1er au 11 août 1875: Compte rendu des séances* (Paris: Imprimerie de E. Martinet, 1875), v.

45. See Charles-Gabriel de Morlet, *Topographie des Gaules: Notice sur les voies romaines du Département du Bas-Rhin* (Strasbourg: Imprimerie de Veuve Berger-Levrault, 1861), 6. See also Krzysztof Pomian, "Francs et Gaulois," in *Les lieux de mémoire*, ed. Pierre Nora, vol. 3 (Paris: Gallimard, 1992).

46. See Herman Lebovics, *True France: The Wars over Cultural Identity, 1900–1945* (Ithaca: Cornell University Press, 1992).

47. See Maurice Barrès, *Les déracinés* (Paris: Charpentier, 1897).

48. F. Dollinger, "A quelle race apartiennent les Alsaciens?" *Revue alsacienne illustrée* 5 (1903): 1–9.

49. E. Blind, "Histoire anthropologique de l'Alsace," *Revue alsacienne illustrée* 5 (1903): 94. For a reiteration of this argument, see Louis Batiffol, *L'Alsace est française par ses origines, sa race, son passé* (Paris: Ernest Flammarion, 1919), 31.

50. Blind, "Histoire anthropologique de l'Alsace," 95.

51. Ibid.

52. See Archives départementales du Bas-Rhin, 27 AL 698.

53. The mapping project was formally announced at the German Association for History and Antiquity meeting at Sigmarien in 1891. See Dr. Friedrich von Thudichum, *Historisch-statistiche Grundkarten: Denkschrift* (Tübingen: Verlag der H. Laupp'schen Buchhandlung, 1892).

54. "Über ein neues Verfahren zur Herstellung historischer Karten: Vortrag des Herrn. Professors Dr. F. von Thudichum aus Tuebingen in der ersten Sitzung der 37. Generalversammlung der deutschen Geschichts und Alterthumsvereine zu Sigmaringen," in ibid.

55. Resolutions from the General Meeting of the German History and Antiquity Association, 1 September 1891, Sigmarien, in ibid.

56. Ibid.

57. See Metz Museum photographic collection, KM 436.

58. See *Travaux du Comité d'études: L'Alsace-Lorraine et la frontière du nord-est*, vol. 1, 1918, Archives départementales du Bas-Rhin, 4K16.

59. For a discussion of the ethnographic map of Romania used in territorial negotiations at Versailles, see Gilles Palsky, "Emmanuel de Martonne and the Ethnological Cartography of Central Europe (1917–1920)," *Imago Mundi* 54 (2002): 111–19.

60. On the 1923 population exchange between Greece and Turkey, see Onur Yildirim, *Diplomacy and Displacement: Reconstructing the Turco-Greek Exchange of Populations, 1922–1934* (New York: Routledge: 2006); Renée Hirschon, *Crossing the Aegean: An Appraisal of the 1923 Compulsory Population Exchange between Greece and Turkey* (New York: Berghahn Books, 2002).

61. Rogers Brubaker uses the legal distinctions between France's *jus solis* and Germany's *jus sanguinis* definition of citizenship to argue that France is fundamentally a political nation while Germany is an ethnocultural nation. See Rogers Brubaker, *Citizenship and Nationhood in France and Germany* (Cambridge, Mass.: Harvard University Press, 1992). For another excellent study on the subject, see Brian Singer, "Cultural versus Contractual Nations: Rethinking Their Opposition," *History and Theory* 35, no.3 (1996): 309–37.

62. See David Allen Harvey, "Lost Children or Enemy Aliens? Classifying the Population of Alsace after the First World War," *Journal of Contemporary History* 34, no. 4 (October 1999): 537–54; Laird Boswell, "From Liberation to Purge Trials in the 'Mythic Provinces': The Reconfiguration of Identities in Alsace and Lorraine, 1918–1920," *French Historical*

Studies 23 (Winter 2000): 129–62. For a comparative European perspective, see Tara Zahra, "The Minority Problem: National Classification in the French and Czechoslovak Borderlands," *Contemporary European History* 17 (May 2008): 137–65.

63. See Boswell, "From Liberation to Purge Trials," 3.

64. Ibid., 13.

65. See report from the Direction de police de Strasbourg, 9 August 1919, Archives départementales du Bas-Rhin, 121 AL 952.

66. See letter from Professor Erich Jung to the Oberkommissar in Strassburg, 7 April 1919, Archives départementales du Bas-Rhin, 121 AL 952.

67. See, for example, Elly Heuss-Knapp, *Souvenirs d'une allemande de Strasbourg, 1881–1934*, trans. Jean-Yves Mariotte (Strasbourg: Éditions Oberlin, 1996).

68. Statistics cited in Städtischen Statistischen Amt, *Verwaltungsbericht der Stadt Strassburg für die Zeit von 1870 bis 1888/89* (Strassburg: Elsässische Druckerei und Verlagsanstalt vorm. G. Fischbach, 1895), 17.

69. Georg Wolfram and Werner Gley, eds., *Elsass-Lothringer Atlas: Landes-kunde, Geschichte, Kultur und Wirtschaft Elsass-Lothringens* (Frankfurt: Selbstverlag des Elsass-Lothringischen Instituts), 1931. For a French critique of the atlas, see Lucien Gallois, "Un atlas d'Alsace-Lorraine," *Annales de géographie* 1 (September 1932): 518–23.

70. For an excellent overview of cartography's role in German revanchist politics during the interwar years and World War II, see Guntram Henrik Herb, *Under the Map of Germany: Nationalism and Propaganda, 1918–1945* (New York: Routledge, 1996). See also Astrid Mehmel, "Deutsche Revisionspolitik in der Geographie nach dem Ersten Weltkrieg," *Geographische Rundschau* 9 (September 1995): 498–505.

71. For an updated edition of Febvre's original 1935 text, see Lucien Febvre, *Le Rhin: Histoire, mythes et réalités* (Paris: Perrin, 1997).

72. Ibid., 40.

73. Ibid., 49.

Chapter 3

1. The dialects were the historical descendants of languages inherited from the Frankish and Alemannic invasions of the fifth century. Alsatian (*Alsacien*, or *Elsässisch*) is closely tied to Swiss and southern German dialects, while Franconian (*Mosellan*, or *Francisque*) is related to Rhineland German. See, for example, Marie-Noële Denis, "Le dialecte alsacien: État des lieux," *Ethnologie française* 33, no. 3 (July 2003): 363–71.

2. Daniel Nordman, the leading historian of French borders, rightly argues that visual representations of linguistic territories in the nineteenth century did not reflect social reality, but rather made arguments, and even claims, about territory. "Not only was language inscribed into the territory," he writes, "but it identified [the territory] more and more with the nation." See Daniel Nordman, *Frontières de la France: De l'espace au territoire, XVI–XIXe siècle* (Paris: Gallimard, 1998), 497. In recent years, the Czech-German language border in particular has attracted the attention of scholars of nationalism. See Pieter Judson, *Guardians of the Nation: Activists on the Language Frontiers of Imperial Austria* (Cambridge, Mass.: Harvard University Press, 2006); Mark Cornwall, "The Struggle on the Czech-German Language Border, 1880–1940," *English Historical Review* 109, no. 433 (September 1994): 914–51.

3. Historians of France and Germany have for decades concentrated on the role of language in national-identity formation—analyzing how the French and German languages were cataloged, promoted, and schooled—but little scholarly attention has been paid to the dramatic "territorialization" of language in the modern European landscape. For general studies on the relationship between language and national-identity formation in Germany and France, see Michael Towson, *Mother-Tongue and Fatherland: Language and Politics in Germany* (New York: St. Martin's Press, 1992); Andreas Gardt, ed., *Sprachgeschichte als Kulturgeschichte* (Berlin: De Gruyter, 1999); Sussan Ameri, *Die Deutschnationale*

Sprachbewegung im Wilhelminischen Reich (New York: P. Lang, 1991); Tuska Benes, *In Babel's Shadow: Language, Philology and the Nation in Nineteenth-Century Germany* (Detroit: Wayne State University Press, 2008); Michel de Certeau, ed., *Une politique de la langue: La révolution française et les patois* (Paris: Gallimard, 2002); Jacques-Philippe Saint-Gérard, ed., *Mutations et sclérose: La langue française, 1789–1840* (Stuttgart: F. Steiner, 1993); Gérald Antoine and Robert Martin, eds., *Histoire de la langue française, 1914–1945* (Paris: CNRS Éditions, 1995); Bernard Cerquiglini, ed., *Le français dans tous ses états* (Paris: Flammarion, 2002).

4. "A cartographic interpretation of silences on a map," writes J. B. Harley, "departs . . . from the premise that silence elucidates and is likely to be as culturally specific as any other aspect of the map's language." See J. B. Harley, "Silences and Secrecy: The Hidden Agenda of Cartography in Early Modern Europe," chap. 3 in *The New Nature of Maps: Essays in the History of Cartography* (Baltimore: Johns Hopkins University Press, 2001), 86.

5. On counter-mapping, see Nancy Lee Peluso, "Whose Woods Are These? Counter-mapping Forest Territories in Kalimantan, Indonesia," *Antipode* 27, no. 4 (October 1995): 383–406.

6. The Edict of Villers-Cottêrets (1539) required French as the language of use for administrative documents in the French kingdom.

7. See David Bell, "National Language and the Revolutionary Crucible," chap. 6 in *The Cult of the Nation in France: Inventing Nationalism, 1680–1800* (Cambridge, Mass.: Harvard University Press, 2001). For a discussion of eighteenth-century and revolutionary language politics in Alsace, see David Bell, "Nation-Building and Cultural Particularism in Eighteenth-Century France: The Case of Alsace," *Eighteenth-Century Studies* 21, no. 4 (Summer 1988): 472–90. See also De Certeau, *Politique de la langue*.

8. Bell, *Cult of the Nation*, 175.

9. Paul Lévy's two-volume study offers a magisterial survey of language politics in Alsace and Lorraine from the Roman period until the twentieth century. I would like to thank David Bell for referring me to this text. Paul Lévy, *Histoire linguistique d'Alsace et de Lorraine*, 2 vols. (Strasbourg: Imprimerie Alsacienne, 1929), 2:22.

10. Lévy, *Histoire linguistique*, 2:24.

11. Napoleon Bonaparte, who spoke French with a Corsican accent, famously said of the Alsatian generals who served him: "Leave these brave men to their own language; their swords fight in French [*ils sabrent en français*]." Quoted in Lévy, *Histoire linguistique*, 2:72.

12. Coquebert de Montbret's study returned 350 versions from seventy-four departments. See Nordman, *Frontières de France*, 488–89. See also T. Bulot, "L'enquête de Coquebert de Montbret et la glottopolitique de l'Empire français," *Romanischen Philologie* 2, no. 89 (1989): 287–92.

13. For an excellent discussion of Coquebert de Montbret's map production and career with the French state, see Isabelle Laboulais, "Reading a Vision of Space: The Geographical Map Collection of Charles-Étienne Coquebert de Montbret (1755–1831)," *Imago Mundi* 56, no. 1 (2004): 48–66; Isabelle Laboulais, "Modalités de construction: Un savoir cartographique et mobilisation des réseaux de correspondants; Le cas des ego-documents de Charles-Étienne Coquebert de Montbret (1755–1831)," in *Nouvelles approches des espaces et réseaux rationnels*, ed. Pierre-Yves Beaurepaire and Dominique Taurisson (Montpellier: Université Montpellier III, 2003), 97–118; Isabelle Laboulais, *Lectures et pratiques de l'espace, l'itinéraire de Coquebert de Montbret (1755–1831), savant et grand commis d'État* (Paris: Honoré Champion, 1999).

14. See the correspondence and maps relating to Coquebert de Montbret's study in the file entitled "Patois de la France: Limites de la langue française," NAF 5913, Bibliothèque nationale de France.

15. See letters exchanged between the French Statistical Bureau and the subprefect of the Arrondissement of Deseuments, 4 September 1806, NAF 5913, Bibliothèque nationale de France. A discussion of the complexity of gathering statistics in nineteenth-century

French provinces can be found in M. N. Bourguet, *Déchiffrer la France: La statistique départementale à l'époque napoléonienne* (Paris: Éditions des archives contemporaines, 1989). See also Jean-Claude Perrot, *L'âge d'or de la statistique régionale française (an IV–1804)* (Paris: Société des études Robespierristes, 1977).

16. Feminist geographers have frequently referred to the map's perspective as a "god's-eye view," for its domineering, all-seeing vision of land. See, for example, Mei-Po Kwan, "Feminist Visualization: Re-envisioning GIS as a Method in Feminist Geographic Research," *Annals of the Association of American Geographers* 92, no. 4 (December 2002): 647.

17. For a discussion of the invention of "thematic maps" displaying social information in early nineteenth-century France, see Josef W. Konvitz, *Cartography in France, 1660–1848: Science, Engineering, and Statecraft* (Chicago: University of Chicago Press, 1987), 127–28. See also Gilles Palsky, *Des chiffres et des cartes: La cartographie quantitative au XIXe siècle* (Paris: Comité des travaux historiques et scientifiques, 1996).

18. Quoted in Lévy, *Histoire linguistique*, 2:41.

19. See Jacob Grimm and Wilhelm Grimm, *Märchen der Brüder Grimm* (Winheim: Beltz und Gelberg, 1995). For a discussion of the relationship between German society and German language politics before national unification, see Ruth Sanders, *German: Biography of a Language* (Oxford: Oxford University Press, 2010).

20. Maps of Germany (or Germania) had existed before the advent of nineteenth-century nationalism, but they did not have rigorous standards for locating precise borders and relied on indistinct dotted lines. See Helmut Walser Smith's discussion of "senses of the nation before nationalism" in *The Continuities of German History: Nation, Religion, and Race across the Long Nineteenth Century* (Cambridge: Cambridge University Press, 2008), 39–73.

21. Thongchai Winichakul argues that maps are often focused on a future spatial order. See Thongchai Winichakul, *Siam Mapped: A History of the Geo-body of a Nation* (Honolulu: University of Hawaii Press, 1994) 130.

22. See the preface to Bernhardi's map for a discussion of his research methods. See also Morgane Labbé's discussion of Bernhardi's 1844 map in "Les frontières de la nation allemande dans l'espace de la carte, du tableau statistique et de la narration," in *L'espace de l'Allemagne au XIXe siècle: Frontières, centres et question nationale*, ed. Catherine Maurer (Strasbourg: Presses universitaires de Strasbourg, 2007), 4.

23. Heinrich Nabert's map *Über Sprachgrenzen insonderheit die deutsch-französischen in den Jahren 1844–1847* was published in *Berghaus' Physikalischer Atlas* 2, no. 9 (Gotha: Berghaus, 1848).

24. Bernhard Struck discusses the political effects of the 1849 version of Bernhardi's language map on Prussia's Polish-speaking eastern borderlands. See Bernhard Struck, "Farben, Sprachen, Territorien: Die deutsch-polnische Grenzregion auf Karten des 19. Jahrhunderts," in *Kartenwelten: Der Raum und seine Repräsentation in der Neuzeit*, ed. Christof Dipper and Ute Schneider (Darmstadt: Primus, 2006), 187–89.

25. Historians and social scientists have recently focused on the impact of language censuses on the development of national identities in Central and Eastern Europe. See Morgane Labbé, "Le projet d'une statistique des nationalités discuté dans les sessions du Congrès International de Statistique (1853–1876)," in *Démographie et politique*, ed. Hervé le Bras et al. (Dijon: Éditions Universitaires de Dijon, 1997); Dominique Arel and David I. Kertzer, eds., *Census and Identity: The Politics of Race, Ethnicity and Language in National Censuses* (Cambridge: Cambridge University Press, 2002). For Prussia, see Morgane Labbé, "Dénombrer les nationalités en Prusse au XIXe siècle: Entre pratique d'administration locale et connaissance statistique de la population," *Annales de Démographie Historique* 1 (2003): 39–61. For the Austro-Hungarian Empire, see Judson, *Guardians of the Nation*; and Morgane Labbé, "La carte ethnographique de l'empire autrichien: La multinationalité dans l'ordre des choses," *Monde des cartes* 180 (2004): 71–84. For Russia, see Francine Hirsch, *Empire of Nations: Ethnographic Knowledge and the Making of the Soviet Union* (Ithaca: Cornell Univer-

sity Press, 2005); and Juliette Cadiot, *Le laboratoire imperial: Russie-URSS, 1870–1940* (Paris: CNRS Editions, 2007).

26. For an excellent biography of Böckh, see a collection of readings from his funeral: *Zur Erinnerung an Richard Böckh: Geheimen Regierungsrat und Professor der Statistik an der Universität Berlin: Reden bei der Trauerfeier am 9. Dezember 1907: Lebenslag und Schriftenübersicht*. Yale Microfilm Collection.

27. Richard Böckh, *Der Deutschen Volkszahl und Sprachgebiet in den europäischen Staaten* (Berlin: Verlag von J. Guttentag, 1869), 5.

28. Böckh, *Deutschen Volkszahl und Sprachgebiet*, 5.

29. See Labbé, "Carte ethnographique," 1.

30. Böckh argued that French language laws in Alsace and German-speaking Lorraine were the equivalent of "a declaration of war against the German nation." Böckh, *Deutschen Volkszahl und Sprachgebiet*, 17.

31. For a discussion of this organization, see Judson, *Guardians of the Nation*, 16.

32. See chart found in Böckh, *Deutschen Volkszahl und Sprachgebiet*, annex.

33. Lévy, *Histoire linguistique*, 2:331.

34. See Christian Jacob, *The Sovereign Map: Theoretical Approaches in Cartography throughout History*, trans. Tom Conley (Chicago: University of Chicago Press, 2006), 207.

35. In 1871, Kiepert and Böckh also collaborated on a historical map of Alsace-Lorraine from the seventeenth and eighteenth centuries. They labeled the map with German toponyms dating from the Holy Roman Empire: the endings of town names included "bach," "berg," "heim," "hofen," "hausen," "ingen" and "weiler." See Richard Böckh and Heinrich Kiepert, *Historische Karte von Elsass und Lothringen zur Uebersicht der territorialen Veränderungen im 17. und 18. Jahrhundert* (Berlin: Verlag von Dietrich Reimer, 1871).

36. Ferdinand Mentz, "Die Ortsnamenverdeutschung in Elsass-Lothringen," *Zeitschrift des Allgemeinen Deutschen Sprachvereins* 31, no. 1 (January 1916): 4–8.

37. See H. Kiepert, *Bemerkungen und Berichtigungen zu dem Verzeichniss der Gemeinden des Depr. Deutsch Lothringen geordnet nach Sprachgebieten*, H. Kiepert to Von Moeller, 12 October 1871, Archives départementales du Bas-Rhin, 159 AL 120.

38. Letter from the Mayor of Vaudoncourt to the German government, 12 February 1872, Archives départementales du Bas-Rhin, 159 AL 120.

39. Statistic cited in Lévy, *Histoire linguistique*, 2:398. The difference between the language boundary and the western boundary of the German-occupied parts of Alsace during the Franco-Prussian War (as of 30 August 1870) can be seen in Augustus Petermann, "Das General-Gouvernement Elsass und die deutsch-französische Sprachgrenze," *Pettermanns Geographische Mittheilungen* 16 (1870): plate 22.

40. A noted exception to this "practical" approach of moving language borders to suit the German Empire's defensive needs was the geographer Friedrich Ratzel's continued insistence on the superficiality of Germany's "political borders" compared to its enduring "natural borders." See map entitled "Deutsches Reichs- und Sprachgrenze," in Friedrich Ratzel, *Politische Geographie* (Munich: Verlag von R. Oldenbourg, 1897), 521.

41. Heinrich Kiepert, "Die Sprachgrenze in Elsass-Lothringen," *Zeitschrift der Gesellschaft für Erdkunde zu Berlin* 9 (1874): 307.

42. Ibid.

43. See also Constant This, *Die Deutsch-Französische Sprachgrenze in Lothringen* (Strassburg: J. H. Ed. Heitz, 1887).

44. Constant This, *Die Deutsch-Französische Sprachgrenze im Elsass* (Strassburg: J. H. Ed. Heitz, 1888), 1–2.

45. See L. Liebich, "Esquisse d'une carte linguistique de l'Alsace," *Revue d'Alsace* 2, no. 2 (1861). The fact that a priest was the first person to map the Alsatian dialect was not surprising. During the French Revolution, priests were some of the staunchest defenders of the Germanic-based Alsatian dialect against revolutionaries who wanted to purge the

border region of both its backward religiosity and its backward language. See Joseph F. Byrnes, "The Relationship of Religious Practice to Linguistic Culture: Language, Religion and Education in Alsace and Roussillon, 1860–1890," *Church History* 68, no. 3 (September 1999): 598–626.

46. See Liebich, "Esquisse d'une carte linguistique," 1.

47. The Alsatian dialect was the focal point of a renaissance in Alsatian culture during the nineteenth century. Beginning in the 1830s, middle- and upper-class Alsatians organized a number of scientific and historical societies that focused on preserving and elevating the status of Alsace's particular regional culture. A patois that had once been scorned by educated regional elites as a lower-class jargon was transformed into the preferred medium of communication for the Alsatian bourgeoisie's plays, poems, and literature. The development of regionalism was not unique to Alsace; across nineteenth-century Europe, educated elites revived their regional cultures. For an excellent discussion of Alsatian regionalism, see Christopher Fischer, *Alsace to the Alsatians? Visions and Divisions of Alsatian Regionalism, 1870–1939* (New York: Berghahn, 2010).

48. Liebich to Eduard von Moeller, Ober-Präsident of Alsace-Lorraine, 2 July 1873, Archives départementales du Bas-Rhin, 159 AL 546.

49. Liebich to the Ober-Präsident of Alsace-Lorraine, 28 April 1873, Archives départementales du Bas-Rhin, 159 AL 546.

50. During the 1860s, France was the sole country in Europe that did not include language as a category on its national census. In 1863, the French government did conduct an internal language survey through school reports. For an account of the survey, see Eugen Weber, *Peasants into Frenchmen: The Modernization of Rural France, 1870–1914* (Stanford: Stanford University Press, 1976), 67–69.

51. Letter from Liebich to the governor-general of Alsace-Lorraine, 28 April 1873, Archives départementales du Bas-Rhin, 159 AL 546.

52. Liebich's survey was entitled *Fragebogen und Schema zur Verfertigung einer Elsässer Grammatik nebst Sprachkarte für Elsass und Deutsch-Lothringen.* See ibid.

53. Letter from Liebich to the Ober-Präsident, 2 November 1874, Archives départementales du Bas-Rhin, 159 AL 546.

54. Ibid.

55. Liebich, *Entwurf einer Sprachkarte des MittelElsasses*, 1876, Archives départementales du Bas-Rhin,159 AL 546.

56. Liebich to the governor-general of Alsace-Loraine, 28 April 1873, Archives départementales du Bas-Rhin, 159 AL 546.

57. For an overview of the evolution of dialect mapping methods, see Wilbert Herringa and John Nerbonne, "Dialect Areas and Dialect Continua," *Language Variation and Change* 13 (2001): 375–400.

58. See Peluso, "Whose Woods Are These?"

59. See letter from Liebich to Von Moeller, 2 July 1873, Archives départementales du Bas-Rhin, 159 AL 546.

60. H. Lienhart and E. Martin, *Wörterbuch der elsässischen Mundarten*, vol. 1 (Strassburg: Verlag von Karl J. Trübner, 1899).

61. See Mentz, "Die Ortsnamenverdeutschung in Elsass-Lothringen," *Zeitschrift des Allgemeinen Deutschen Sprachvereins* 31, no. 1 (January 1916): 4.

62. Peter Paulin, "Die Ortsnamenverdeutschung in Elsass-Lothringen," *Petermanns Geographische Mittheilungen*, no. 62 (1916): 121.

63. Ibid., 121.

64. See Archives départementales du Bas-Rhin, 159 AL 136.

65. See Stephen Harp's chapter "Schooling in the First World War," in *Learning to Be Loyal: Primary Schooling as Nation Building in Alsace and Lorraine, 1850–1940* (DeKalb: Northern Illinois University Press, 1998), 160–82.

66. See Bruno Cabanes, *La victoire endeuillée: La sortie de la guerre des soldats français, 1918–1920* (Paris: Seuil, 2004). See also Stéphane Audoin-Rouzeau and Annette Becker, *14–18: Rétrouver la guerre* (Paris: Gallimard, 2000).

67. Albert de Dietrich, *Alsaciens corrigeons notre accent!* (Paris: Berger-Levrault, 1917), 4.

68. Ibid., x–xi.

69. See Gilles Palsky, "Emmanuel de Martonne and the Ethnological Cartography of Central Europe (1917–1920)," *Imago Mundi* 54 (2002): 111–19.

70. See Archives départementales du Bas-Rhin, 98 AL 614.

71. Office régionale de stastitique, 1931 census, Archives départementales du Bas-Rhin, 98 AL 614.

72. See Samuel Huston Goodfellow, *Between the Swastika and the Cross of Lorraine: Fascisms in Interwar Alsace* (DeKalb: Northern Illinois University Press, 1999).

73. Karl Roos, *Die Fremdwörter in den Elsässischen Mundarten: Ein Beitrage zur Elsässischen Dialektforschung* (Strassburg: Universitäts-Buchdrückerei von J. H. Ed. Heitz, 1903).

Chapter 4

1. In his groundbreaking study on space, geographer Yi-Fu Tuan argues that the human desire for centeredness is reflected in how people orient maps. "People everywhere," he writes, "tend to regard their homeland as the 'middle place' or the center of the world." See Yi-Fu Tuan, *Space and Place: The Perspective of Experience* (Minneapolis: University of Minnesota Press, 2001), 38.

2. In 1850, 63% of Alsatians lived in communities with fewer than two thousand residents. In 1900, the figure stood at 50%, and by 1954, it was 41%. Statistics cited in Alfred Wahl and Jean-Claude Richez, *L'Alsace entre France et Allemagne, 1850–1950* (Paris: Hachette, 1994), 18.

3. Kären Wigen's study on regional maps from Japan's Nagano Prefecture offers a fruitful comparative example of how regional maps served as intermediaries between national and regional identities. The critical role of regional mapmakers in Japan's territorial unification demonstrates, in her words, that "the passage to modernity—and the route to national identity—led not around the province but through it." Kären Wigen, *A Malleable Map: Geographies of Restoration in Central Japan, 1600–1912* (Berkeley: University of California Press, 2010), 19.

4. Historians have recently explored the idea of *Heimat* as instrumental to bottom-up nation building in imperial Germany. For a discussion of the various strands of regionalism that developed in Alsace, see Christopher Fischer, *Alsace to the Alsatians? Visions and Divisions of Alsatian Regionalism, 1870–1939* (New York: Berghahn, 2010). According to Celia Applegate, the idea of *Heimat* "bridged the gap between national aspiration and provincial reality." See the introduction to Celia Applegate, *A Nation of Provincials: The German Idea of Heimat* (Berkeley: University of California Press, 1990). Further, Alon Confino argues that *Heimat* helped Germans to "devise a common denominator between their intimate, immediate, and real local place and the distant, abstract, not-less-real national world." See Alon Confino, *The Nation as a Local Metaphor: Württemberg, Imperial Germany, and National Memory, 1871–1918* (Chapel Hill: University of North Carolina Press, 1997), 4.

5. Historians of modern France have likewise shifted toward an understanding of nation building that involved celebrating, rather than crushing, regional diversity. See Jean-François Chanet, *L'École républicaine et les petites patries* (Paris: Aubier, 1996); Caroline Ford, *Creating the Nation in Provincial France: Religion and Political Identity in Brittany* (Princeton: Princeton University Press, 1993); Anne-Marie Thiesse, *Ils apprenaient la France: L'exaltation des régions dans le discours patriotique* (Paris: Éditions de la Maison des Sciences de l'Homme, 1997).

6. Alon Confino's book is a notable exception. In his study on the intersection of regional and national identity in Württemberg, Confino turned to several hundred *Heimat*

images to analyze how ordinary provincial Germans visualized their nation. He connected the growing strength of regionalism in turn-of-the-century Württemberg to the dissemination of new kinds of local images that encouraged provincials to conceptualize their collective identity through their physical surroundings. See his chapter "The Nation in the Mind" in Confino, *Nation as a Local Metaphor*, 158–209.

7. At the same time, this chapter is careful to point out Alsace-Lorraine's special historical role within both French and German states. Unlike the other regional states within the German Empire, which enjoyed substantial federal powers, Alsace-Lorraine was designated as a *Reichsland*, or "imperial territory," by the rest of the German states, meaning that it was subject to direct rule from Berlin until it was finally granted a constitution and recognized as an autonomous federal state in 1911. Moreover, "Alsace-Lorraine" was itself a historical anomaly as a region. Created by the German Empire in 1871 out of Alsace and parts of Lorraine, both centuries-old French provinces, the region was composed of two populations with different cultural backgrounds. Therefore, when regionalism spread across Europe during the turn of the century, two regionalist movements existed side by side in Alsace-Lorraine, both Alsatian and Lorrainer forms.

8. Chorography refers to a practice of regional description that was popular in early modern Europe. It was a form of mapping that emphasized the particularism of individual places rather than Euclidean space. In his study on the Scottish chorographer Timothy Pont, Ian Cunningham argues that chorography was a wide-ranging example of the "geographical arts" that emphasized local topography, human production, family histories, and small-scale environments. See Ian C. Cunningham, *The Nation Survey'd: Essays on Late Sixteenth-Century Scotland as Depicted by Timothy Pont* (East Lothian: Tuckwell Press, 2001), 141–42; See also Lesley B. Cormack, "Good Fences Make Good Neighbors: Geography as Self-Definition in Early Modern England," *Isis* 82, no. 4 (December 1991): 639–61. According to Kären Wigen, a chorographic study was characterized by "its middling size, mediating between the intimate horizons of everyday life and the expansive world beyond." See Wigen, *Malleable Map*, 14–15.

9. Scholars of European cartography have argued that the framing of local space was critical to building relationships between local populations and their larger national or imperial community. "Maps," writes Valerie Kivelson, "are mediators between the individual and the political structure—particularly for translating power relations to the local level and for people to understand their place within a collectivity." See Valerie Kivelson, *Cartographies of Tsardom: The Land and Its Meanings in Seventeenth-Century Russia* (Ithaca: Cornell University Press, 2006), 11.

10. On print culture and the "vernacularization of space" through mass-produced maps, see Marcia Yonemoto, *Mapping Early Modern Japan: Space, Place and Culture in the Tokugawa Period (1603–1868)* (Berkeley: University of California Press, 2003), 14.

11. Kirsten Belgum's study of a popular German magazine, *Die Gartenlaube*, offers a useful model for exploring how mass-produced visual images created a familiar German geography in the minds of the public. See Kirsten Belgum, *Popularizing the Nation: Audience, Representation, and the Production of Identity in "Die Gartenlaube," 1853–1900* (Lincoln: University of Nebraska Press, 1998).

12. Stéphane Gerson describes the French concept of the *pays* as "more a collection of images and feelings than a bounded area." See Stéphane Gerson, *The Pride of Place: Local Memories and Political Culture in Nineteenth-Century France* (Ithaca: Cornell University Press, 2003), 5. Alon Confino argues that new kinds of local images encouraged nineteenth-century German provincials to conceptualize their collective identity in their physical surroundings. See his chapter "The Nation in the Mind," in Confino, *Nation as a Local Metaphor*, 158–209.

13. Recent scholars of modern Mexico and its borderlands have used the concept of a "fugitive landscape" to explore power struggles between states and local societies. See

Raymond B. Craib, *Cartographic Mexico: A History of State Fixations and Fugitive Landscapes* (Durham: Duke University Press, 2004); Samuel Truett, *Fugitive Landscapes: The Forgotten History of the U.S.-Mexico Borderland* (New Haven: Yale University Press, 2008).

14. Christopher Fischer calls the region "an intermediate territorial level between state and locality." See Fischer, *Alsace to the Alsatians?*, 9.

15. See, for example, David Hopkin, "Identity in a Divided Province: The Folklorists of Lorraine, 1860–1960," *French Historical Studies* 23, no. 4 (2000): 639–82.

16. My methodology for analyzing cadastral maps draws on Valerie Kivelson's argument that maps can have an "expressive function" in addition to their stated "instrumental function" and can therefore serve as a medium to express utopian and idealized understandings of community. See Kivelson, *Cartographies of Tsardom*, 5.

17. The French cadastre began in 1807, when Napoleon decided to reorganize France's labyrinthine tax system. He called upon departmental and communal authorities to measure more than one hundred million parcels of land in France and to draw a plan for each French commune that indicated the precise layout of these parcels. Surveyors created cadastral plans that distinguished the various property sections in the commune, labeling each section with the name of its owner. The information that each the village plan provided would help to determine each owner's tax obligations. According to Napoleon's system, property that contained building structures was taxed at a higher rate than "bare" (*nu*) territory. See the Departmental Archives of the Bas-Rhin's historical overview of the cadastre and guidebook to the archive's cadastre holdings. See also "Carte représentant l'ancienneté des plans cadastraux des départements du Bas-Rhin, du Haut-Rhin et de la Moselle," 1933, Archives départementales du Bas-Rhin, 4K21.

18. The German cadastre differed from the French: it taxed the size of landed property as well as the goods produced from exploiting the land. Properties without buildings were assessed according to the net revenue of the terrain: income earned from vineyards, gardens, fields, pastures, woods, and mines. Built properties, including houses, barns, cellars, mills, and factories, were evaluated based on their dimensions. See ibid.

19. This collection of cadastral maps (labeled *plans d'assemblage* or *Uebersichtskarte*) is available for consultation at the Departmental Archives of the Bas-Rhin in map collection Plan/2P.

20. E. P. Thompson has demonstrated that village boundaries could become part of a community's collective memory through annual or regular perambulation. See E. P. Thompson, *Customs in Common* (New York: New Press, 1991), 98.

21. I would like to thank Georges Bischoff, professor of history at the University of Strasbourg, for his help in analyzing this image.

22. See a 1915 discussion regarding cadastral maps in communes and their classrooms. Archives départementales du Bas-Rhin, 105 AL 2138.

23. See letter from the Directeur des contributions directes et du cadastre d'Alsace et de Lorraine to the Rector d'Académie d'Alsace et de Lorraine in Strasbourg regarding the use of cadastral maps in schools, 8 April 1921, Archives départementales du Bas-Rhin, 105 AL 2138.

24. Ibid.

25. I am indebted to Alon Confino's chapter on the *Heimat* imagery. Confino, *Nation as a Local Metaphor*, 168–69.

26. The term "Heimats" is an anglicized version of the plural for "Heimat." In German, there is no plural version of the word "Heimat." Ibid., 125.

27. G. Wickersheimer, *Geographie von Elsass-Lothringen bearbeitet von einem Mitgliede der Strassburger Lehrer-Conferenz* (Strassburg: Druck und Verlag von J. H. Eduard Heitz, 1872), 1–2.

28. See Jean-Jacques Rousseau, *Émile* (Paris: Hachette: 1880).

29. Wickersheimer, *Geographie von Elsass-Lothringen*, 32.

30. Ibid, 35.

31. For a discussion of the bourgeoisie's role in *Heimatkunde*, see Applegate, *Nation of Provincials*.

32. See, for example, Johan Ludwig Algermissen, *Georg Langs Mittelschul-Atlas für Elsass-Lothringen mit besonderer Berücksichtigung der Heimats- und Vaterlandskunde* (Gebweiler: Boltzeschen Buchhandlung, 1890).

33. See lesson on the *Heimatort* in Georg Weick, *Heimatkunde (Heimat und Kreis): Ein Hilfsbüchlein für den ersten Geographieunterricht* (Zabern: Druck und Verlag der Schulbuchhandlung H. Fuchs, 1894).

34. Ibid.

35. Joseph Lefftz, *Das Volkslied im Elsass*, vol. 1 (Colmar: Editions Alsatia, 1966), 273.

36. For similar songs celebrating Alsatian pride in the local landscape, see "Heimat und Fremde," "Wo ist ein Land so schön?," "Das Elsass ist das beste Land," and "'S Elsass, unser Ländel," in ibid.

37. Stephen Harp quotes one German curriculum that discussed how folk songs "gave direct expression to the emotions, natural senses, and love of the fatherland." See Harp, *Learning to Be Loyal*, 113.

38. Lefftz, *Volkslied im Elsass*, 7.

39. Harp, *Learning to Be Loyal*, 114.

40. In 1790, French revolutionaries called for the elimination of the state's patchwork of provinces, each with its own special privileges and relationships to the king, in favor of eighty-three equally sized and jurisdictionally identical administrative units called departments (*départements*), innocuously named after French rivers and mountains. The revolutionaries' project of territorial *découpage* was intimately linked to their beliefs about the proper relationship between the particular and the universal in a democratic republic. By wiping the Old Regime provinces off the map of France, they hoped to destroy the noble power structure for which they had served as a base, replacing it with a system of government with equal rights for all citizens. For an image of France with its new departmental divisions, see C. E. Delamarche, *Le Royaume de France Divisé en 83 Départemens suivant les Décrets de l'Assemblée Nationale des 15 Janvier, 16 et 26 Fevrier 1790, Patentés par le Roi des François le 4 Mars de la même année*, 1790, Bibliothèque nationale de France, Ge C 1985. For the definitive study on the destruction of the Old Regime provinces and the making of the departmental system, see Marie-Vic Ozouf-Marignier, *La formation des départements: La représentation du territoire français à la fin du 18e siècle* (Paris: Éditions de l'école des hautes études en sciences sociales, 1989). See also J. E. Gerock, *La formation des départements du Haut-Rhin et du Bas-Rhin en 1789* (Thann: Imprimerie du Journal de Thann, 1925).

41. Th. Scharf, *Geographie für die Volksschulen von Elsass-Lothringen* (Metz: Schulbuchhandlung von Wwe. Alcan, 1876), 1.

42. Weick, *Heimatkunde*, 7.

43. See *Verlagskatalog der J. Boltzeschen Buchhandlung O. H. Gebweiler, 1870–1910* (Leipzig: Buch- u. Kunstdruckerei Breitkopf & Haertel, 1910).

44. Ibid.

45. David Gugerli and Daniel Speich, *Topografien der Nation: Politik, kartographische Ordnung und Landschaft im 19. Jahrhundert* (Zurich: Chronos, 2002) 12–13.

46. Algermissen, *Georg Langs Mittelschul-Atlas*, 2.

47. On railroads as a unifying force, see Weber, *Peasants into Frenchmen: The Modernization of Rural France, 1870–1914* (Stanford: Stanford University Press, 1976).

48. G. Moser and G. Kaufmann, *Geographische Faustzeichnung als Grundlage für einen methodischen Unterricht in der Geographie* (Strassburg: R. Schultz u. Comp., 1875). The use of this method in Alsatian schools predated the German annexation. During the Second Empire, Alsatian schools used Parisian-printed cartographic instruction manuals, authored by a national educational society, that emphasized departmental landmarks. See Aug. Braud,

Cartographie élémentaire des écoles: Méthode nouvelle et progressive de géographie pratique mise à la portée des enfants, Cahier cartographique, no. 67 (Paris: Dezobry, F. Tandou et Co, n.d.).

49. Algermissen, *Georg Langs Mittelschul-Atlas*, 1.

50. Martin Kunz, *Das Bild in der Blindenschule* (Kiel: Druck von Schmidt und Klaunig, 1891), 5.

51. Martin Kunz, *Blinden Atlas*, 1891, Bibliothèque municipale de Strasbourg.

52. See guidebook to the 2006 exposition at the Strasbourg municipal archives: "Les relations franco-allemandes à travers les atlas scolaires allemands et français, 27 février–21 avril 2006."

53. François Pétry, "Invention du paysage et identité aux XIXème et XXème siècles en Alsace," *Revue d'Alsace* 131 (2005): 314.

54. Georges Bischoff discusses the separation between Protestant and Catholic popular imagery in Alsace (including baptismal and marriage certificates) during the eighteenth and early nineteenth centuries, and the subsequent regionalization and laicization of popular images at the turn of the century. See Georges Bischoff, "L'invention de l'Alsace," *Saisons d'Alsace* 119 (1993): 34–69. For a microhistory of religious relations in an Alsatian village, see Rebecca McCoy, "The Culture of Accommodation: Religion, Language and Politics in an Alsatian Community, 1648–1870" (Ph.D. diss., University of North Carolina, Chapel Hill, 1992). For a discussion of the changing place of Jews in Alsatian society, see Paula E. Hyman, *The Emancipation of the Jews of Alsace: Acculturation and Tradition in the Nineteenth Century* (New Haven: Yale University Press, 1992).

55. Quoted from Hans Haug in Pétry, "Invention du paysage," 285.

56. Buhr and Buzon, *Geographie für Elsass-Lothringischen Schulen* (Metz: Druck und Verlag von Paul Even, 1902).

57. E. Rudolph, *Landeskunde des Reichslandes Elsass-Lothringen* (Breslau: Königliche Universitäts- und Verlagsbuchhandlung, 1907).

58. For a discussion of Alsatian housing styles and regional identity, see Jean-Claude Vigato, "L'architecture régionaliste de 1900 à 1930," *Revue d'Alsace* 131 (2005): 165–88.

59. Pétry, "Invention du paysage."

60. See Georges Bischoff, "Provocation, patriotisme et poésie: L'Alsace de Hansi," *Historiens et géographes* 86 (February 1995): 213–20.

61. The 1912 exhibition's official catalog, *Internationalen Fremden-und-Reise-Verkehrs-Ausstellung*, can be found in Archives départementales du Bas-Rhin, 87 AL 3262.

62. See Rachel Chrastil, *Organizing for War: France, 1870–1914* (Baton Rouge: Louisiana State University Press, 2010).

63. Weber, *Peasants into Frenchmen*.

64. In a process that Thiesse calls the "patrimonialization of local identities," French schoolteachers taught their students that the national and the local were "perfectly complementary"; see Thiesse, *Ils apprenaient la France*, 7. Jean-François Chanet writes that teachers did not bring a "cultural genocide" to the provinces, but rather, like the Germans, they worked with existing local cultures to cultivate affection for the larger country; see Chanet, *École républicaine*. Caroline Ford discusses the role of the local Breton elite, particularly members of the clergy, in adapting French nationalism to the particularities of Brittany; see Ford, *Creating the Nation in Provincial France*. For a recent discussion of cultural particularity and national unity in modern France, see Patrick Young, *Enacting Brittany: Tourism and Culture in Provincial France, 1871–1939* (Surrey: Ashgate, 2012).

65. In her study on the gardens of Versailles, Chandra Mukerji argues that the French king's seventeenth-century gardens created a certain model for French style and taste that would eventually disseminate to the rest of France. See Chandra Mukerji, *Territorial Ambitions and the Gardens of Versailles* (Cambridge: Cambridge University Press, 1997), 98. Another centralist vision of French territory that contrasted with Vidal de la Blache's view was articulated by Jules Michelet in his *Tableau de la France*, published in 1833. Unlike Jules

Michelet, Vidal believed that all of France's regions were on an equal footing, and that Paris did not rise above the rest. For a detailed discussion of Vidal de la Blache's school of geography, its influences, and its competitors, see Jean-Yves Guiomar, "Vidal de la Blache's Geography of France," in *Realms of Memory: The Construction of the French Past*, ed. Pierre Nora, trans. Arthur Goldhammer, vol. 1 (New York: Columbia University Press, 1997), 192.

66. Paul Vidal de la Blache, *Tableau de la géographie de France* (Paris: Hachette, 1903).

67. For a discussion of Vidal's concept of the *genius loci*, see Guiomar, "Vidal de la Blache's Geography of France," 204.

68. Maurice Agulhon, *Marianne into Battle: Republican Imagery and Symbolism in France, 1789–1880*, trans. Janet Lloyd (Cambridge: Cambridge University Press, 1981), 4–5.

69. Joan Landes, *Visualizing the Nation: Gender, Representation and Revolution in Eighteenth-Century France* (Ithaca: Cornell University Press, 2001).

70. Adolphe Braun and Christian Kempf, *Alsace photographiée en 1859* (Obernai: Gyss, 2003).

71. G. Bruno, *Le tour de la France par deux enfants* (Paris: Librairie Classique Eugène Belin, 1905), 318.

72. Jacques Ozouf and Mona Ozouf, "*Le tour de la France par deux enfants*: The Little Red Book of the Republic," in Nora, *Realms of Memory*, 131.

73. Bruno, *Le tour de la France*, preface; hereafter, page numbers are indicated in the text.

74. Ozouf and Ozouf, "*Le Tour de la France*," 129.

75. For an analysis of the maps and illustrations in Fouillée's book, see Kory Olson, "Creating Map Readers: Republican Geographic and Cartographic Discourse in G. Bruno's (Augustine Fouillée) 1905 Le Tour de la France par deux enfants," *Modern and Contemporary France* 19, no. 1 (February 2011): 37–51.

76. Ozouf and Ozouf, "*Le Tour de la France*," 22.

77. Bernard Lightman argues that "the golden age of female popularization of science in the second half of the nineteenth century was . . . a by-product of the lack of options for women who were fascinated by the world of nature." Bernard Lightman, *Victorian Popularizers of Science: Designing Nature for New Audiences* (Chicago: University of Chicago Press, 2007), 488.

78. In her best-selling geography reader, originally published in 1906, Selma Lagerlöf narrates the story of a young boy who learns about his national homeland of Sweden from the back of a flying goose. For an English-language edition, see Selma Lagerlöf, *The Wonderful Adventure of Nils* (New York: Doubleday, 1913). For an in-depth discussion of Emma Willard, author of the first classroom geography atlas for American schools, see Susan Schulten, *Mapping the Nation: History and Cartography in Nineteenth-Century America* (Chicago: University of Chicago Press, 2012), 18–40.

79. Archives départementales du Bas-Rhin, 121 AL 1093.

80. Ibid.

81. *Petite géographie de l'Alsace et la Lorraine à l'usage des écoles* (Strasbourg: A. Viz et Co. Librairies-Éditeurs, 1919).

82. Alain Corbin, *Village Bells: Sound and Meaning in the Nineteenth-Century French Countryside*, trans. Martin Thom (Cambridge, Mass.: Harvard University Press, 1994).

83. On homogenized church imagery in popular German print, see Confino, *Nation as a Local Metaphor*, 163. In Alsace, many village churches did in fact have a history of mixed Protestant and Catholic use. See Claude Muller and Bernard Vogler, *Catholiques et protestants en Alsace: Le simultaneum de 1802 à 1892* (Strasbourg: Librairie ISTRA, 1983).

84. *Ésprit du clocher* can be defined as parochialism or a *limitation de vues* (limitation of views). The belief that villagers were reluctant to look outside their own small world reinforced the Third Republic idea that the countryside could be modernized by expanding villagers' views of the outside world through education and visual tools such as maps.

85. *Petite géographie de l'Alsace et la Lorraine*.

86. Mona Siegel argues that the state allowed French schoolteachers the flexibility to choose their own instructional materials and design their own lesson plans during the interwar years. See Mona Siegel, *The Moral Disarmament of France: Education, Pacifism, and Patriotism, 1914–1940* (Cambridge: Cambridge University Press, 2004).

87. Hennigé's text was specifically designed for the Upper-Alsatian village of Fréland (formerly Urbach under German rule). Fr. E. Hennigé. *L'Alsace: Géographie locale et régionale à l'usage des écoles primaires des lycées et des collèges avec modèle d'une monographie géographique du lieu de domicile, carte en couleurs, plans et croquis* (Colmar: Société alsacienne d'édition Alsatia, 1920).

88. Selection from Alphonse Daudet, "Contes choisis," in *Lectures alsaciennes: Géographie, histoire, biographies,* ed. Christian Pfister (Paris: Librairie Armand Colin, 1919).

89. The cement canal, designed by the engineer René Koechlin, was to run parallel to the Rhine, diverting the water from its original riverbed. Eight hydroelectric stations were to be built along the canal, generating energy from a series of strategically placed artificial waterfalls. The purpose of the canal was therefore twofold: to produce electrical power and to serve as a transportation link between Basel, Mulhouse, Colmar, and Strasbourg. One of the largest rectification projects undertaken on the Upper Rhine since the Tulla Rectification Projects (1817–76), the plan was funded by the French government, local Alsatian banks, and the Société Énergie Électrique du Rhin. See Charles Béliard, *Le Grand Canal d'Alsace: Voie navigable, source d'énergie* (Nancy: Imprimerie Berger-Levrault, 1926).

Chapter 5

1. See Goethe's accounts of his travels through the Vosges in Johann Wolfgang von Goethe, *Aus meinem Leben, Dichtung und Wahrheit* (Berlin: H. Seemann Nachfilger, 1900). Late nineteenth-century French writers such as Émile Erckmann and Alexandre Chatrian used the symbolism of "la ligne bleu des Vosges" in their works focusing on the loss of Alsace-Lorraine to Germany. See, for example, E. Erckmann and A. Chatrian, *L'ami Fritz* (Paris: Hachette, 1882).

2. Describing the power of tourist maps, Henri Lefebvre writes: "Maps that show 'beauty spots' and historical sites and monuments to the accompaniment of an appropriate rhetoric aim to mystify in fairly obvious ways. This kind of map designates places where a ravenous consumption picks over the last remnants of nature and of the past in search of whatever nourishment may be obtained from the *signs* of anything historical or original. If the maps and guides are to be believed, a veritable feast of authenticity awaits the tourist." See Henri Lefebvre, *The Production of Space*, trans. D. Nicholson-Smith (Oxford: Basil Blackwell, 1991), 84.

3. See Michel de Certeau, *The Practice of Everyday Life*, trans. Steven E. Rendall (Berkeley: University of California Press, 1984), 118–21. In his study on mapping Australia, Paul Carter describes the practice of walking from place to place as a "peripatetic narrative" that allows space to "unfold horizontally." See Paul Carter, *The Road to Botany Bay: An Exploration of Landscape and History* (Minneapolis: University of Minnesota Press, 2010), 7. In Latin America, the walking of boundaries was also an important ritual. The boundaries on some Latin American maps were lined with a string of footprints. "As their connection to perambulation suggests," argues Barbara Mundy, "it is the human presence that defines space, both through naming, and through movement." See Barbara Mundy, *The Mapping of New Spain* (Chicago: University of Chicago Press, 1996), 111.

4. On perambulation around the bounds of parishes in England, see E. P. Thompson, *Customs in Common* (New York: New Press, 1991), 98.

5. Simon Schama, *Landscape and Memory* (New York: Knopf, 1995), 15.

6. David Blackbourn, *The Conquest of Nature: Water, Landscape, and the Making of Modern Germany* (Cambridge, Mass.: Harvard University Press, 2006), 15.

7. Charles Harrison, "The Effects of Landscape" in W. J. T. Mitchell, ed. *Landscape and*

Power (Chicago: University of Chicago Press, 2002), 212. See also James Duncan and David Ley, eds. *Place/Culture/Representation* (London: Routledge, 1993); Denis Cosgrove and Stephen Daniels, eds., *The Iconography of Landscape: Essays on the Symbolic Representation, Design, and Use of Past Environments* (Cambridge: Cambridge University Press, 1988); Stephen Daniels, *Fields of Vision: Landscape Imagery and National Identity in England and the United States* (Cambridge: Polity Press, 1993).

8. See chap. 3, "Culture on the Ground: The World Perceived through the Feet," in Tim Ingold, *Being Alive: Essays on Movement, Knowledge and Description* (London: Routledge, 2011), 33–50. French historian Alain Corbin likewise defines landscape as a mode of perceiving space that relies not only on sight, but on all of the senses, including touch, smell, and sound. See Alain Corbin, *L'homme dans le paysage, entretien avec Jean Lebrun* (Paris: Les éditions textuel, 2001), 9.

9. Ingold argues: "Locomotion, not cognition, must be the starting point for the study of perceptual activity. Or, more strictly, cognition should not be *set off* from locomotion, along the lines of a division between head and heels, since walking is itself a form of circumambulatory knowing." See Ingold, *Being Alive*, 46.

10. See Chandra Mukerji, *Territorial Ambitions and the Gardens of Versailles* (Cambridge: Cambridge University Press, 1997).

11. W. J. T. Mitchell advocates using the term "landscape" as a verb. See W. J. T. Mitchell, "Introduction," in *Landscape and Power*, ed. W. J. T. Mitchell (Chicago: University of Chicago Press, 2002), 1–2; Simon Schama, furthermore, notes that the word "landscape" comes from the Dutch term "landschap," which literally means the "shaping" of land. See Schama, *Landscape and Memory*, 10.

12. The Vogesenclub published accounts of these early voyages in its club journal. See, for example, "Vogesenreisen in alter Zeit III: Wanderungen von Joh. Andreas Silbermann," *Mittheilungen aus dem Vogesenclub*, vol. 38 (1904).

13. Corbin, *Homme dans le paysage*, 104. Corbin discusses the first tourist book that popularized the idea of the picturesque and romantic voyage in France. See Isidore Baron Taylor and Charles Nodier, *Voyages pittoresques et romantiques dans l'ancienne France* (Paris: Didot, 1833).

14. See a contemporary reproduction of the album in Adolphe Braun and Christian Kempf, *L'Alsace photographiée en 1859* (Obernai: Gyss, 2003), 14.

15. Stéphane Gerson explores how local elites in provincial France developed "cults of local memories" during the nineteenth century, helping to create a new form of French national identity that reached outside Paris and incorporated the regions' unique histories and cultures. Civic participation through associational life was the key to developing this new idea of territorial identity. See Stéphane Gerson, *The Pride of Place: Local Memories and Political Culture in Nineteenth-Century France* (Ithaca: Cornell University Press, 2003).

16. L. Levrault, Th. de Morville, and X. Mossmann, *Musée pittoresque et historique de l'Alsace: Dessins et illustrations par J. Rothmuller* (Colmar: J. Rothmuller, Editeur, 1863).

17. Here I use Pierre Nora's definition of a "site of memory" as "any signifying entity, of a material or ideal kind, which has through human will or the work of time become a symbolic element of the memorial patrimony of a given community." See Pierre Nora, "Comment écrire l'histoire de France?," in *Les lieux de mémoire*, ed. Pierre Nora, vol. 3 (Paris: Gallimard, 1992), 11.

18. Levault, de Morville, and Mossmann, *Musée pittoresque*, 18.

19. Ibid., 20. For a similar guide to Ribeauvillé, see Frédéric Piton, *Promenades en Alsace: Monographies historiques, archéologiques et statistiques; Ribeauvillé et ses environs* (Strasbourg: Chez l'auteur, 1856).

20. Ibid., 20.

21. This image can be found in the 1865 edition of the *Atlas national illustré*, Archives départementales du Bas-Rhin, 2K27.

22. Frédéric Kirschleger, "Guide du botantiste herborisateur et touriste à travers les plaines de l'Alsace et les montages des Vosges," in *Flore d'Alsace et des contrées limitrophes*, vol. 3 (Strasbourg: Imprimerie d'Ed. Huder, 1862), 342.

23. Ibid., 323–24.

24. In a similar collaboration between Alsatians and Badeners, medical doctor Robert Aimé used international scientific research to publish the first French-language description of the mineral baths in Alsace, including Wattwiller and Bad Niederbronn. Aimé thanked Professor Bunsen from Heidelberg (inventor of the Bunsen burner) for sending him a manuscript and letting him copy his analysis, as well as the government of Baden, which made information on the region's mineral baths available to him. See Robert Aimé, *Guide du médecin et du touriste aux bains de la vallée du Rhin, de la Forêt-Noire et des Vosges: Avec plusieurs analyses inédites de M. Bunsen, professeur de chimie à l'Université de Heidelberg* (Paris: L. Hachette et Comp., 1857), viii–ix.

25. Otto Bechstein and Hans Luthmer, *Bericht über die Thätigkeit des Vogesenclubs in den ersten 25 Jahren seines Bestehens* (Strassburg: Heitz und Mündel, 1897), 7.

26. For further discussion of the relationship between hiking clubs and conservation in modern Germany, see John Alexander Williams, *Turning to Nature in Germany: Hiking, Nudism, and Conservation* (Stanford: Stanford University Press, 2007); Thomas Lekan, *Imagining the Nation in Nature: Landscape Preservation and German Identity, 1885–1945* (Cambridge, Mass.: Harvard University Press, 2004).

27. See Hans Luthmer, "Der Vogesenclub von 1872–1918," in *Das Reichsland Elsass-Lothringen, 1871–1918*, vol. 3, ed. Georg Wolfram (Frankfurt: Selbstverlag des Elsass-Lothringen Instituts, 1934), 502.

28. Bechstein and Luthmer, *Bericht über die Thätigkeit des Vogesenclubs*, 7.

29. See Luthmer, "Vogesenclub," 499.

30. See Julius Euting's introduction to G. M. Eckert, *Bilder aus dem Elsass* (Heidelberg: Verlag von Fr. Bassermann, 1874).

31. See image 22 in ibid.

32. Curt Mündel, *Die Vogesen: Reisehandbuch für Elsass-Lothringen und angrenzende Gebiete* (Strassburg: Verlag von Karl J. Trübner, 1881).

33. See Euting's instructions for mapmaking in *Mittheilungen aus dem Vogesenclub*, vol. 5 (June 1876).

34. For an overview of the Vogesenclub's map production, see Otto Bechstein, "Das Kartenwerk des Vogesenclubs: Ein Rück- und Überblick," *Mittheilungen aus dem Vogesenclub*, vol. 43 (October 1910).

35. See Stephan Oettermann, *The Panorama: History of a Mass Medium* (New York: Zone Books, 1997), 49.

36. Ibid., 7.

37. According to Bernard Comment, maps' flattened views created a "landscape that appeared neither inhabited nor alive, but disconnected from any figure or subject that gave it a sense of place." See Bernard Comment, *The Painted Panorama* (London: Reaktion Books, 1999), 86.

38. See Corbin, *Homme dans le paysage*, 21; Mary Louise Pratt also discusses the nineteenth-century European desire for heroic promontory views in her study on the colonial appropriation of land in Africa and Latin America. See Mary Louise Pratt, *Imperial Eyes: Travel Writing and Transculturation* (New York: Routledge, 1992), 202.

39. For an analysis of the politics surrounding the monuments on Vosges crests, see Jean-Marc Dreyfus, "Eine Grenze in Ruinen: Zur Symbolik der Gipfel in den Vogesen," in *Wiedergewonnene Geschichte: Zur Aneignung von Vergangenheit in den Zwischenräumen Mitteleuropas*, ed. Peter Oliver Loew, Christian Pletzing, and Thomas Serrier (Wiesbaden: Harrassowitz Verlag, 2006), 363–82.

40. J. Naeher, *Panorama vom Donon im Elsass* (Strassburg im Elsass: Verlag von J. H. Ed. Hetiz, 1888).

41. See description of his Donon panorama in *Südwestdeutsche Touristen-Zeitung* 1, no. 2 (April 1895).

42. Otto Bechstein, *Der Donon und seine Denkmäler* (Strassburg: J. H. Ed. Heitz, 1891).

43. Gustave Fraipont, *Les montagnes de France: Les Vosges* (Paris: Librairie Renouard, 1894), 403.

44. See the image from *Mittheilungen aus dem Vogesenclub*, no. 32 (31 October 1899).

45. Charles Victor de Hohenlohe-Schillingsfürst was governor of Alsace-Lorraine from 1885 to 1894. Prince Hermann Hohenlohe-Langenburg succeeded him as prefect of Alsace-Lorraine from 1894 to 1907.

46. Letter from the Ministry of Alsace-Lorraine granting assistance of six thousand marks for the construction of the tower, 30 May 1895, Archives départementales du Bas-Rhin, 314 D 41.

47. For an account of the inauguration of the Hohenlohe Tower, see "Die Einweihung des Hohenlohe-Turms auf dem Hochfelde am 2 Oktober 1898," *Mittheilungen aus dem Vogesenclub*, vol. 32 (October 1898). See also Auguste Gyss, "L'historique de la Tour du Champ du Feu," *Les Vosges*, no. 4 (1984).

48. See *Nos Vosges: Revue mensuelle du Touring-Club Vosgien* 2, no. 11 (August–October 1913).

49. Luthmer, "Vogesenclub," 500.

50. *Nos Vosges: Revue mensuelle du Touring-Club Vosgien* 2, no. 11 (August–October 1913).

51. *Nos Vosges: Revue mensuelle du Touring-Club Vosgien* 2, no. 10 (May–June 1913).

52. See François Pétry, "Invention du paysage et identité aux XIXème et XXème siècles en Alsace," *Revue d'Alsace* 131 (2005): 314.

53. *Nos Vosges: Revue mensuelle du Touring-Club Vosgien* 1. no. 8 (January 1912).

54. See Flori Frei, "Heimat," *Der Vogesenwanderer* 1, no. 1 (May 1920).

55. Touring-Club de France to the Commissaire Général de la République in Strasbourg, 1 April 1919, Archives départementales du Bas-Rhin, 121 AL 97.

56. *Compte rendu: Assemblée générale du 1er juin 1919 du Club Vosgien*, Archives départementales du Bas-Rhin, 314 D 41.

57. *Bulletin du Club Vosgien*, no. 3 (March 1922).

58. *Bulletin du Club Vosgien*, no. 2 (February 1922).

59. "Rapport du Président à l'Assemblée générale à Sélestat," *Bulletin du Club Vosgien*, no. 8 (August 1925).

60. Ibid.

61. On battlefield tourism after the First World War, see David Lloyd, *Battlefield Tourism: Pilgrimage and Commemoration of the Great War in Britain, Australia and Canada, 1919–1939* (Oxford: Berg, 1998).

62. Henri Martin, *Le Vieil Armand, 1915* (Paris: Payot, 1936), 9.

63. See Comité de l'Hartmannswillerkopf, "Vue panoramique du champ de bataille de l'Hartmannswillerkopf," 26 March 1922, Archives départementales du Bas-Rhin, 121 AL 1091.

64. Maréchal Pétain, "Le Monument National de l'Hartmannswillerkopf," 1927, Archives départementales du Bas-Rhin, 98 AL 626.

65. For practical reasons, most of France's "martyred landscapes" were restored after World War I. See Hugh Clout, *After the Ruins: Restoring the Countryside of Northern France after the Great War* (Exeter: University of Exeter Press, 1996).

66. On postwar battlefield tourism in France, see Stephen Harp, *Marketing Michelin: Advertising and Cultural Identity in Twentieth-Century France* (Baltimore: Johns Hopkins Univer-

sity Press, 2001), 99–123. See also Daniel Sherman, *The Construction of Memory in Interwar France* (Chicago: University of Chicago Press, 2001.

67. See the list of criteria to be met by the monument. Comité de l'Hartmannswillerkopf et Comité de la Reconnaissance Alsacienne-Lorraine, "Monument de l'Hartmannswillerkopf," Archives départementales du Bas-Rhin, 121 AL 1091.

68. Ibid.

69. "Le Monument du Hartmannswillerkopf," *Bulletin du Club Vosgien* 1, no. 2 (October 1921).

70. Ibid.

71. See F. J. Hahn and J. L. Pinol, "La mobilité d'une grande ville: Strasbourg 1870 à 1940," *Annales de démographie historique* 1 (1995): 197–210.

Chapter 6

1. For an excellent study on the shifting symbolism of Strasbourg as a postnational "European" border city after World War II, see John Western, *Cosmopolitan Europe: A Strasbourg Self-Portrait* (London: Ashgate, 2012). Other recent discussions of European border cities include Ed Taverne and Cor Wagenaar, "Border Cities: Contested Identities of the European City," *European Review* 13, no. 2 (2005): 201–6; Andrew Herscher, "Urban Formations of Difference: Borders and Cities in Post-1989 Europe," *European Review* 13, no. 2 (2005): 251–60.

2. "Social geography" refers to the idea that the spaces that you move through can "reveal who you are, what you do, and where you come from." See David Harvey, *Paris: Capital of Modernity* (New York: Routledge, 2005), 40.

3. This chapter focuses on the evolution of Strasbourg's modern cityscape, leaving a discussion of the city's social and political history to other authors. See, for example, Georges Livet and Francis Rapp, eds., *Histoire de Strasbourg* (Toulouse: Privat, 1987). For a longer analysis, see vols. 3 and 4 of their earlier series: Georges Livet and Francis Rapp, eds., *Histoire de Strasbourg des origines à nos jours*, 4 vols. (Strasbourg: Éditions des Dernières Nouvelles de Strasbourg, 1980–81), vols. 3–4.

4. European philosophers and social critics have long debated the influences of urban space on collective mentalities. See Henri Lefebvre, *The Production of Space*, trans. D. Nicholson-Smith (Oxford: Basil Blackwell, 1991). For a recent comparative study of European urban theory, see Jenny Bavidge, *Theorists of the City: Walter Benjamin, Henri Lefebvre and Michel de Certeau* (New York: Routledge, 2009).

5. Johannes Striedbeck, *Argentina*, 1760, Bibliothèque nationale de France, Ge D 16981. For a theoretical analysis of the image, see the chapter entitled "The City in Its Map and Portrait," in Louis Marin, *On Representation*, trans. Catherine Porter (Stanford: Stanford University Press, 2001), 202–18.

6. Yi-Fu Tuan explains the difference between the profile view, which people can see with their own eyes, and the artificial bird's-eye view. "Human beings live on the ground and see trees and houses from the side," he writes. "The bird's eye view is not ours, unless we climb a tall mountain or fly in an airplane." See Yi-Fu Tuan, *Space and Place: The Perspective of Experience* (Minneapolis: University of Minnesota Press, 2001), 27.

7. Louis Marin makes this claim about the surveyor's vantage point in Marin, *On Representation*, 211.

8. For the larger context of nineteenth-century French urban development, see John Merriman, *The Margins of City Life* (New York: Oxford University Press, 1991).

9. Lucien Febvre discusses the military origin of the word *frontière*, which derives from the war term "front." See Lucien Febvre, "*Frontière*: The Word and the Concept," in *A New Kind of History: From the Writings of Febvre*, ed. Peter Burke (London: Routledge, 1973).

10. Letter from Strasbourg in Victor Hugo, *Le Rhin: Lettres à un ami*, vol. 2 (Paris: Imprimerie Nationale, 1985), 126.

11. Frédéric Piton, *Strasbourg illustré ou panorama pittoresque, historique et statistique de Strasbourg et de ses environs* (Strasbourg: Chez l'auteur, rue du Temple Neuf, 15, et chez les principaux librairies, 1855).

12. Piton, *Strasbourg illustré*, preface.

13. Catherine Bertho Lavenir argues that French guidebooks from the 1860s, such as the *Guide Joanne Bade et la Forêt Noire,* did not focus on a strict national frontier and concentrated on satisfying the tastes of the "romantic traveler." See Catherine Bertho Lavenir, "Strasbourg, un guide à main (1863–1930)," *Revue d'Alsace* 131 (1995): 221–40.

14. In July of 1870, when Napoleon III declared war on the German states over a disputed succession to the Spanish crown, Bismarck used the opportunity to mobilize German nationalism and cement a political union between northern and southern Germany. Though it would be remembered as the "Franco-Prussian" War, troops from all the German states participated. Setting a pattern to be repeated in the twentieth century, this first violent encounter between France and the German Empire would be fought on Alsatian and Lorrainer soil, in cities such as Metz, Wissembourg, and Sedan, where the strength of the unified German forces and the weakness of the French counterattack contributed to a string of defeats for Napoleon III. For an excellent account of the siege of Strasbourg, see Rachel Chrastil, *The Siege of Strasbourg* (Cambridge, Mass.: Harvard University Press, 2014).

15. Gustave Doré, *Le Rhin allemand*, Staatsbibliothek zu Berlin, Preussischer Kulturbesitz, Handschriftenabteilung, YC 8007 g.

16. A letter from Oscar Berger-Levrault printed in the newspaper provided instructions for how to diffuse a shell safely and bury it. See *Courrier du Bas-Rhin*, 13 September 1870.

17. Gustave Fischbach, *Le siège et le bombardement de Strasbourg* (Strasbourg: Chez les principaux librairies, 1870), 165. For another personal account of the siege, see Jules-Edouard Dufrenoy, *Journal du siège de Strasbourg: Texte inédit présenté par Emmanuel Amougou* (Paris: L'Harmattan, 2004). For a firsthand account of siege events, published posthumously, see also Frédéric Piton, *Siège de Strasbourg: Journal d'un assiégé* (Paris: Charles Schlaeber, 1900).

18. Marc Bonnefroy, *Strasbourg en 1870: Notes et impressions d'un officier pendant le siège* (Paris: Librairie Alsacienne-Lorraine, 1911), 37.

19. Fischbach, *Siège et le bombardement*, 57–58.

20. See notice in the *Courrier du Bas-Rhin*, 5 October 1870.

21. Fischbach, *Siège et le bombardement*, 58.

22. Winter recorded damages caused to the cathedral in Charles Winter, "Rapport au Maire Küss des dégats causés à la cathédrale, 1871," Archives de Strasbourg.

23. M. E. Müntz, "Les monuments d'art détruits à Strasbourg," *Gazette des Beaux Arts*, April 1872.

24. See notice in the *Courrier du Bas-Rhin*, 6 October 1870.

25. The German Empire spent a total of 1,018,724 marks to repair the damage from the siege. See "Krieg 1870–71, Belagerung und Beschiessung der Stadt," in Städtischen Statistischen Amt, *Verwaltungsbericht der Stadt Strassburg für die Zeit von 1870 bis 1888/89* (Strassburg: Elsässische Druckerei und Verlagsanstalt vorm. G. Fischbach, 1895), 543.

26. See Angéla Kerdilès Weiler, *Limites urbaines de Strasbourg: Évolution et mutation* (Strasbourg: Publications de la société savante d'Alsace, 2005).

27. See Klaus Nohlen's comprehensive book on the construction of Strasbourg's New City after 1871: *Construire une capitale: Strasbourg impérial de 1870 à 1918; Les bâtiments officiels de la Place Impériale* (Strasbourg: Publications de la société savante d'Alsace, 1997), 37.

28. See Architekten und Ingenieur-Verein für Elsass-Lothringen, *Strassburg und seine Bauten* (Strassburg: Verlag von Karl J. Truebner, 1894).

29. Song recorded in *Kommers zu Ehren des Bürgermeister der Stadt Strassburg Herrn Otto Back aus Anlass seines 70 jährigen Geburtstages veranstaltet von den städtischen Beamten am 20ten Oktober 1904*, Bibliothèque nationale et universitaire de Strasbourg, M.22030.

30. See *Beckmann's Fuehrer durch Strassburg und Umgebung* (Stuttgart: Verlag von Klemm und Beckmann, 1905), 100–102.

31. See the account of the emperor's visit to Alsace-Lorraine in October 1895 in Archives de Strasbourg, 282 M W68.

32. Klaus Nohlen argues that the medieval structure of the Old City was left intact to remind citizens of Strasbourg's history as a free imperial city of the Holy Roman Empire. See Klaus Nohlen, "Das Bild der Stadt Strassburg zur Reichslandzeit: Historischer Kern versus Neustadt," *Revue d'Alsace* 131 (2005): 139–63. See also Nohlen's description of the sight lines and axes of alignment in Nohlen, *Construire une capitale*, 43.

33. See Johann Gottfried Conrath, *Plan der Stadt Strassburg und Ihrer Erweiterung*, 1878. Bibliothèque nationale et universitaire de Strasbourg, M.Carte.10629.

34. Image found in *Kommers zu Ehren*.

35. Students of German architecture still visit Strasbourg today to study the Wilhelmine building style, since many of the examples from other German cities were destroyed during the bombings of World War II. For a description, see Fritz Beblo, "Die Baukunst in Elsass-Lothringen, 1871–1918," in *Das Reichsland Elsass-Lothringen, 1871–1918*, vol. 3, ed. Georg Wolfram (Frankfurt: Selbstverlag des Elsass-Lothringen Instituts, 1934), 242.

36. See membership listings in *Geschäfts-Ordnung für den Vorstand des Verschönerungs-Vereins für Strassburg und dessen Umgebung*, 5 August 1880, Archives de Strasbourg, 205 MW 32.

37. Catherine Bertho Lavenir divides the evolution of Strasbourg's tourist guidebooks into three periods: the *Guide Joanne Bade et la Forêt Noire* (1863), which did not focus on a strict frontier and concentrated on the romantic traveler; the Baedecker guides from 1880 to 1914, which stressed German patriotic considerations; and the *Michelin Guide Vosges-Alsace-Lorraine*, 1930, which focused on sites of French historical memory. See Bertho Lavenir, "Strasbourg, un guide à la main," 235. For a discussion of the national politics of the tourist gaze, see Rudy Koshar, "What Ought to Be Seen: Tourists' Guidebooks and National Identities in Modern Germany and Europe," *Journal of Contemporary History* 33, no. 3 (July 1998): 323–40.

38. *Beckmann's Fuehrer durch Strassburg*.

39. August Schricker, *Strassburg und Umgebung* (Zürich: J. Laurencie, 1890).

40. Eugen König, *Führer durch Strassburg und die Vogesen* (Strassburg: G. Fischbach, 1896).

41. See "Die XI. Wanderversammlung des Verbandes deutscher Architekten-und-Ingenieur-Vereine zu Strassburg," *Deutsche Bauzeitung*, 14 September 1894.

42. The German population statistic is cited in Stéphane Jonas et al., eds., *Strasbourg, capitale du Reichsland Alsace-Lorraine et sa nouvelle université, 1871–1918* (Strasbourg: Éditions Oberlin, 1994), 10.

43. Livet and Rapp, *Histoire de Strasbourg* (1980–81), 4:219.

44. Elly Heuss-Knapp, *Souvenirs d'une allemande de Strasbourg, 1881–1934*, trans. Jean-Yves Mariotte (Strasbourg: Éditions Oberlin, 1996), 2. Her German memoir was originally published in 1934 under the title *Ausblick vom Münsterturm: Erlebtes aus dem Elsass und dem Reich*.

45. Ibid., 52. For a portrait of the Alsatian elite class, see François Igersheim, *L'Alsace des notables, 1870–1914* (Strasbourg: Budderflade, 1981).

46. Rodolphe Reuss, *Vieux noms et rues nouvelles de Strasbourg: Causeries biographiques d'un flâneur* (Strasbourg: Treuttel et Wuertz, Editeurs, 1883).

47. Adolphe Seyboth, *Ansichten des Alten Strassburg: Fünfzig Tafeln mit erklärendem Text* (Strassburg: J. H. Heitz, 1891), 1–2.

48. Dr. Emerich, "Massnahmen zum Schütze des Ortsbildes gegen Verunstaltung durch Bauausführungen," 1908, Archives de Strasbourg, 152 MW 18.

49. Ibid.

50. See "Bauordnung für die Stadt Strassburg," 8 April 1910, Archives de Strasbourg, 152 MW 18.

51. Supplement to "Bauordnung für die Stadt Strassburg," 23 November 1910, Archives de Strasbourg, 152 MW 21.

52. See Annette Maas, "Stadtplanung und Öffentlichkeit in Strassburg (1870–1918/25): Vom Nationalbewusstsein zur regionalen Identität städtischer Interessengruppen," in *Grenzstadt Strassburg: Stadtplanung, kommunale Wohnungspolitik und Öffentlichkeit, 1870–1940*, ed. Christof Cornelissen et al. (St. Ingbert: Röhrig Universitätsverlag, 1997), 208.

53. See Alon Confino, "The Nation in the Mind," in *The Nation as a Local Metaphor: Württemberg, Imperial Germany, and National Memory, 1871–1918* (Chapel Hill: University of North Carolina Press, 1997).

54. For a detailed description of the parade, see Bruno Cabanes, *La victoire endeuillé: La sortie de guerre des soldats français, 1918–1920* (Paris: Seuil, 2004).

55. See the postcard image *Le général Gouraud passant en revue les troupes françaises, à Strasbourg, le 22 novembre 1918* (Strasbourg: Maison d'art alsacien, 1918), Bibliothèque nationale et universitaire de Strasbourg. See also the documents on the preparation of Strasbourg for the arrival of French troops in Archives de Strasbourg, 282 MW 90.

56. Ibid.

57. See Archives de Strasbourg, 151 MW 80.

58. Ibid.

59. Ibid.

60. *Guide Historique et Artistique de la Ville de Strasbourg*, 1919, Archives de Strasbourg.

61. Guides Michelin, *Strasbourg* (Clermont-Ferrand: Michelin, 1921).

62. "Tribune Publique: Questions de Constructions," *Journal d'Alsace et de Lorraine*, 24 March 1926.

63. Commission des Beaux-Arts, Séances, 1924–1928, Archives de Strasbourg, 152 MW 20.

64. Elisabeth Guévremont, "Les architectes travaillant à Strasbourg durant l'entre-deux-guerres, 1918–1939" (master's thesis, University of Strasbourg, 1997).

65. Service cadastrale de la ville de Strasbourg, *Carte des ports du Rhin de Strasbourg et de Kehl*, 1919. Map archived in 30 AJ 213, Archives nationales.

66. See Archives départementales du Bas-Rhin, 98 AL 378.

67. Commission centrale pour la navigation du Rhin, *La navigation du Rhin* (Strasbourg: Office national de la navigation, 1926), 415.

68. Lucien Febvre, *A Geographical Introduction to History*, trans. E. G. Mountford and J. H. Paxton (New York: Barnes and Noble, 1966), 301. Febvre originally published the book in 1924, when he was a professor of history at the University of Strasbourg.

69. See Archives départementales du Bas-Rhin, 98 AL 378.

70. Ibid.

71. Ibid.

Epilogue

1. The suspension bridge is the work of Parisian architect Marc Mimram. On the architecture of European integration, see Irène Bellier and Thomas M. Wilson, eds., *The Anthropology of the European Union: Building, Imagining and Experiencing the New Europe* (Oxford: Berg, 2000).

2. On the role of Euro-regions in European integration, see Olivier Kramsch and Barbara Hoopers, eds., *Cross-Border Governance in the European Union* (London: Routledge, 2004).

3. On the evolution of the Regio TriRhena, see Susanne Eder and Martin Sandtner, "Common Spirit on the Upper Rhine Valley?," in *Boundaries and Place: European Borderlands in Geographical Context*, ed. David H. Kaplan and Jouni Häkli (New York: Rowman and Littlefield, 2002).

4. For a discussion of the European geographic imagination in the age of European integration, see Alexander P. Murphy, "Rethinking Multi-level Governance in a Changing European Union: Why Metageography and Territoriality Matter," *Geo-Journal* 72, no. 1/2 (2008): 14.

5. Thongchai Winichakul discusses how simple, mass-produced maps of national territory can become "logo maps," or "metasigns," that crystalize national identity. See Winichakul, *Siam Mapped: A History of the Geo-body of a Nation* (Honolulu: University of Hawaii Press, 1994), 138.

6. On educational policy in the European Union, see Brad Blitz, "From Monnet to Delors: Educational Co-operation in the European Union," *Contemporary European History* 12, no. 2 (May 2003): 197–212.

7. For information on the EuroVelo cycling route network and a free cycling map, see "EuroVelo: The European Cycle Route Network," accessed 1 March 2014, http://www.eurovelo.com.

I. Archives and Libraries

Archives de la Ville et de la Communauté urbaine de Strasbourg
Archives départementales du Bas-Rhin, Strasbourg
Archives départementales du Haut-Rhin, Colmar
Archives départementales de Meurthe-et-Moselle, Nancy
Archives nationales, Paris
Bibliothèque nationale, Paris
Bibliothèque nationale et universitaire de Strasbourg
Fonds patrimonial des médiathèques de Strasbourg
Service historique de la défense, Vincennes
Staatsbibliothek zu Berlin–Preussischer Kulturbesitz

II. Newspapers and Journals

Bulletin du Club Vosgien
Courrier du Bas-Rhin
Der Vogesenwanderer
Deutsche Bauzeitung
Elsässer Bilderbogen
Journal d'Alsace et de Lorraine
Mittheilungen aus dem Vogesenclub
Nos Vosges: Revue mensuelle du Touring-Club Vosgien
Petermanns Geographische Mittheilungen
Revue alsacienne illustrée
Revue d'Alsace
Südwestdeutsche Touristen-Zeitung

III. Books, Articles, and Pamphlets

Agulhon, Maurice. *Marianne into Battle: Republican Imagery and Symbolism in France, 1789–1880.* Trans. Janet Lloyd. Cambridge: Cambridge University Press, 1981.

Aimé, Robert. *Guide du médecin et du touriste aux bains de la vallée du Rhin, de la Forêt-Noire et des Vosges: Avec plusieurs analyses inédites de M. Bunsen, professeur de chimie à l'Université de Heidelberg.* Paris: L. Hachette et Comp., 1857.

Akerman, James R. "The Structuring of Political Territory in Early Printed Atlases." *Imago Mundi* 47 (1995): 138–54.

Akerman, James R., and Robert W. Karrow Jr., eds. *Maps: Finding Our Place in the World.* Chicago: University of Chicago Press, 2007.

Algermissen, Johan Ludwig. *Georg Langs Mittelschul-Atlas für Elsass-Lothringen mit besonderer Berücksichtigung der Heimats- und Vaterlandskunde.* Gebweiler: Boltzeschen Buchhandlung, 1890.

Ameri, Sussan. *Die Deutschnationale Sprachbewegung im Wilhelminischen Reich.* New York: P. Lang, 1991.

Anderson, Benedict. *Imagined Communities: Reflections on the Origin and Spread of Nationalism.* London: Verso, 1991.

Anderson, Malcolm. *Frontiers: Territory and State Formation in the Modern World.* Oxford: Polity Press, 1996.

Antoine, Gérald, and Robert Martin, eds. *Histoire de la langue française, 1914–1945.* Paris: CNRS Éditions, 1995.

Applegate, Celia. *A Nation of Provincials: The German Idea of Heimat.* Berkeley: University of California Press, 1990.

Architekten und Ingenieur-Verein für Elsass-Lothringen. *Strassburg und seine Bauten.* Strassburg: Verlag von Karl J. Truebner, 1894.

Arel, Dominique. "Language Categories in Censuses: Backward-or-Forward-Looking?" In *Census and Identity: The Politics of Race, Ethnicity and Language in National Censuses*, ed. Dominique Arel and David I. Kertzer. Cambridge: Cambridge University Press, 2002.

Arel, Dominique, and David I. Kertzer, eds. *Census and Identity: The Politics of Race, Ethnicity, and Language in National Censuses.* Cambridge: Cambridge University Press, 2002.

Audoin-Rouzeau, Stéphane, and Annette Becker. *14–18: Rétrouver la guerre.* Paris: Gallimard, 2000.

Baker, Alan. *Geography and History: Bridging the Divide.* Cambridge: Cambridge University Press, 2003.

Bavidge, Jenny. *Theorists of the City: Walter Benjamin, Henri Lefebvre and Michel de Certeau.* New York: Routledge, 2009.

Barrès, Maurice. *Les déracinés.* Paris: Charpentier, 1897.

Batiffol, Louis. *L'Alsace est française par ses origins, sa race, son passé.* Paris: Ernest Flammarion, 1919.

Baycroft, Timothy. *Culture, Identity and Nationalism: French Flanders in the Nineteenth and Twentieth Centuries.* Suffolk: Royal Historical Society, Boydell Press, 2004.

Bayly, C. A., et al. "AHR Conversation: On Transnational History." *American Historical Review* 111, no. 5 (December 2006): 1441–64.

Beblo, Fritz. "Die Baukunst in Elsass-Lothringen, 1871–1918." In *Das Reichsland Elsass-Lothringen, 1871–1918*, vol. 3, ed. Georg Wolfram. Frankfurt: Selbstverlag des Elsass-Lothringen Instituts, 1934.

Bechstein, Otto. *Der Donon und seine Denkmäler.* Strassburg: J. H. Ed. Heitz, 1891.

———. "Das Kartenwerk des Vogesenclubs: Ein Rück- und Überblick." *Mittheilungen aus dem Vogesenclub*, vol. 43 (October 1910).

Bechstein, Otto, and Hans Luthmer. *Bericht über die Thätigkeit des Vogesenclubs in den ersten 25 Jahren seines Bestehens.* Strassburg: Heitz und Mündel, 1897.

Beckmann's Führer durch Strassburg und Umgebung. Stuttgart: Verlag von Klemm und Beckmann, 1905.

Belgum, Kirsten. *Popularizing the Nation: Audience, Representation, and the Production of Identity in "Die Gartenlaube," 1853–1900.* Lincoln: University of Nebraska Press, 1998.

Béliard, Charles. *Le grand canal d'Alsace, voie navigable, source d'énergie.* Nancy: Imprimerie Berger-Levrault, 1926.

Bell, David. "Nation-Building and Cultural Particularism in Eighteenth-Century France: The Case of Alsace." *Eighteenth-Century Studies* 21, no. 4 (Summer 1988): 472–90.

———. *The Cult of the Nation in France: Inventing Nationalism, 1680–1800.* Cambridge, Mass.: Harvard University Press, 2001.

Bellier, Irène, and Thomas M. Wilson, eds. *An Anthropology of the European Union: Building, Imagining and Experiencing the New Europe.* Oxford: Berg, 2000.

Benes, Tuska. *In Babel's Shadow: Language, Philology, and the Nation in Nineteenth-Century Germany.* Detroit: Wayne State University Press, 2008.

Benjamin, Walter. *Illuminations.* Trans. Harry Zohn. London: Fontana, 1968.

Berghaus, Heinrich. *Physikalischer Atlas.* 2 vols. Gotha: Justus Perthes, 1848.

Bernhardi, Karl. *Sprachkarte von Deutschland.* Kassel: Verlag von J. J. Bohné, 1844.

———. *Die Sprachgrenze zwischen Deutschland und Frankreich.* Kassel: Verlag von A. Freyschmidt, 1871.

Berthaut, Henri. *Les ingénieurs-géographes militaires, 1624–1831.* Vol. 1. Paris: Imprimerie du Service géographique de l'armée, 1902.

———. *Connaissance du terrain et lecture des cartes.* Paris: Imprimerie du Service géographique de l'armée, 1912.

Bertho Lavenir, Catherine. "Strasbourg, un guide à la main (1863–1930)." *Revue d'Alsace* 131 (1995): 221–40.

Biggs, Michael. "Putting the State on the Map: Cartography, Territory, and European State Formation." *Comparative Studies in Society and History* 41, no. 2 (April 1999): 374–405.

Bischoff, Georges. "L'invention de l'Alsace." *Saisons d'Alsace* 119 (1993): 34–69.

———. "Provocation, patriotisme et poésie: L'Alsace de Hansi." *Historiens et géographes* 86 (February 1995): 213–20.

Black, Jeremy. *Maps and Politics.* London: Reaktion Books, 1997.

Blackbourn, David. *The Conquest of Nature: Water, Landscape, and the Making of Modern Germany.* Cambridge, Mass.: Harvard University Press, 2006.

Blind, Edmund. "Histoire anthropologique de l'Alsace." *Revue alsacien illustré* 5 (1903): 89–96.

Blitz, Brad. "From Monnet to Delors: Educational Co-operation in the European Union." *Contemporary European History* 12, no. 2 (May 2003): 197–212.

Bloch, Marc. "Toward a Comparative History of European Societies." In *Enterprise and Secular Change: Readings in Economic History*, ed. Frederic C. Lane and Jelle C. Riemersma. Homewood, Ill.: R. D. Irwin, 1953.

Blum, Alain. "Resistance to Identity Categorization in France." In *Census and Identity: The Politics of Race, Ethnicity and Language in National Censuses*, ed. Dominique Arel and David I. Kertzer. Cambridge: Cambridge University Press, 2002.

Böckh, Richard. *Der Deutschen Volkszahl und Sprachgebiet in den europäischen Staaten.* Berlin: Verlag von J. Guttentag, 1869.

Böckh, Richard, and Heinrich Kiepert. *Historische Karte von Elsass und Lothringen zur Uebersicht der territorialen Veränderungen im 17. und 18. Jahrhundert.* Berlin: Verlag von Dietrich Reimer, 1871.

Boehmer, Georges-Guillaume. *La rive gauche du Rhin, limite de la République française, ou Recueil de plusieurs dissertations dignes des prix proposés par un négociant de la rive gauche du Rhin.* Paris: Chez Desenne, Louvet et Devaux, 1795.

Bonnefroy, Marc. *Strasbourg en 1870: Notes et impressions d'un officier pendant le siège.* Paris: Librairie Alsacienne-Lorraine, 1911.

Boswell, Laird. "From Liberation to Purge Trials in the 'Mythic Provinces': The Reconfiguration of Identities in Alsace and Lorraine, 1918–1920." *French Historical Studies* 23 (Winter 2000): 129–62.

Bourguet, M. N. *Déchiffrer la France: La statistique départementale à l'époque napoléonienne.* Paris: Éditions des archives contemporaines, 1989.

Braud, Aug. *Cartographie élémentaire des écoles: Méthode nouvelle et progressive de géographie pratique mise à la portée des enfants.* Cahier cartographique, no. 67. Paris: Dezobry, F. Tandou et Co., n.d.

Braudel, Fernand. *La Méditerranée et le monde méditerranéen à l'époque de Philippe II.* Paris: Colin, 1949.

———. *L'identité de la France: Espace et histoire.* Paris: Flammarion, 1986.

Braun, Adolphe, and Christian Kempf. *Alsace photographiée en 1859.* Obernai: Gyss, 2003.

Brubaker, Rogers. *Citizenship and Nationhood in France and Germany.* Cambridge, Mass.: Harvard University Press, 1992.

Brubaker, Rogers, and Frederick Cooper. "Beyond 'Identity.'" *Theory and Society* 24, no. 1 (February 2000): 1–47.

Brückner, Martin. *The Geographic Revolution in Early America: Maps, Literacy, and National Identity.* Chapel Hill: University of North Carolina Press, 2006.

Bruno, G. *Le tour de la France par deux enfants.* Paris: Librairie Classique Eugène Belin, 1905.

Budde, Gunilla, et al., eds. *Transnationale Geschichte: Themen, Tendenzen und Theoren.* Göttingen: Vandenhoeck und Ruprecht, 2006.

Buhr and Buzon. *Geographie für Elsass-Lothringischen Schulen.* Metz: Druck und Verlag von Paul Even, 1902.

Buisseret, David, ed. *Monarchs, Ministers, and Maps: The Emergence of Cartography as a Tool of Government in Early Modern Europe.* Chicago: University of Chicago Press, 1992.

Bulot, T. "L'enquête de Coquebert de Montbret et la glottopolitique de l'Empire français." *Romanischen Philologie* 2, no. 89 (1989): 287–92.

Burke, Peter. *Eyewitnessing: The Uses of Images as Historical Evidence.* Ithaca: Cornell University Press, 2001.

Burnett, D. Graham. *Masters of All They Surveyed: Exploration, Geography, and a British El Dorado.* Chicago: University of Chicago Press, 2000.

Byrnes, Joseph F. "The Relationship of Religious Practice to Linguistic Culture: Language, Religion, and Education in Alsace and Roussillon, 1860–1890." *Church History* 68, no. 3 (September 1999): 598–626.

Cabanes, Bruno. *La victoire endeuillée: La sortie de la guerre des soldats français, 1918–1920.* Paris: Seuil, 2004.

Cadiot, Juliette. *Le laboratoire imperial: Russie-URSS, 1870–1940.* Paris: CNRS Editions, 2007.

Caron, Vicki. *Between France and Germany: The Jews of Alsace-Lorraine, 1871–1918.* Stanford: Stanford University Press, 1988.

Carter, Erica et al., eds. *Space and Place: Theories of Identity and Location.* London: Lawrence and Wishart, 1993.

Carter, Paul. *The Road to Botany Bay: An Exploration of Landscape and History.* Minneapolis: University of Minnesota Press, 2010.

Cerquiglini, Bernard, ed. *Le français dans tous ses états.* Paris: Flammarion, 2002.

Chanet, Jean-François. *L'école républicaine et les petites patries.* Paris: Aubier, 1996.

Chester, Lucy P. *Borders and Conflict in South Asia: The Radcliffe Boundary Commission and the Partition of the Punjab.* Manchester: Manchester University Press, 2009.

Chrastil, Rachel. *Organizing for War: France, 1870–1914.* Baton Rouge: LSU Press, 2010.

———. *The Siege of Strasbourg.* Cambridge, Mass.: Harvard University Press, 2014.

Claval, Paul. *Histoire de la géographie française de 1870 à nos jours.* Paris: Nathan, 1998.

Clout, Hugh. *After the Ruins: Restoring the Countryside of Northern France after the Great War.* Exeter: University of Exeter Press, 1996.

Comment, Bernard. *The Painted Panorama.* London: Reaktion Books, 1999.

Commission centrale pour la navigation du Rhin. *La navigation du Rhin.* Strasbourg: Office national de la navigation, 1926.

Confino, Alon. *The Nation as a Local Metaphor: Württemberg, Imperial Germany, and National Memory, 1871–1918*. Chapel Hill: University of North Carolina Press, 1997.

Corbin, Alain. *Village Bells: Sound and Meaning in the Nineteenth-Century French Countryside*. Trans. Martin Thom. Cambridge, Mass.: Harvard University Press, 1994.

———. *L'homme dans le paysage, entretien avec Jean Lebrun*. Paris: Les éditions textuel, 2001.

Cormack, Lesley B. "Good Fences Make Good Neighbors: Geography as Self-Definition in Early Modern England." *Isis* 82, no. 4 (December 1991): 639–61.

Cornelissen, Christof, et al., eds. *Grenzstadt Strassburg: Stadtplannung, kommunale Wohnungspolitik und Öffentlichkeit, 1870–1940*. St. Ingbert: Röhrig Universitätsverlag, 1997.

Cornwall, Mark. "The Struggle on the Czech-German Language Border, 1880–1940." *English Historical Review* 109, no. 433 (September 1994): 914–51.

Cosgrove, Denis, and Stephen Daniels, eds. *The Iconography of Landscape: Essays on the Symbolic Representation, Design, and Use of Past Environments*. Cambridge: Cambridge University Press, 1988.

Craib, Raymond B. "Cartography and Power in the Conquest and Creation of New Spain." *Latin American Research Review* 35, no. 1 (Spring 2000): 7–36.

———. *Cartographic Mexico: A History of State Fixations and Fugitive Landscapes*. Durham: Duke University Press, 2004.

Craig, John E. *Scholarship and Nation Building: The Universities of Strasbourg and Alsatian Society, 1870–1939*. Chicago: University of Chicago Press, 1984.

Crary, Jonathan. *Techniques of the Observer: On Vision and Modernity in the Nineteenth Century*. Cambridge, Mass.: MIT Press, 1990.

Cunningham, Ian C. *The Nation Survey'd: Essays on Late Sixteenth-Century Scotland as Depicted by Timothy Pont*. East Lothian: Tuckwell Press, 2001.

Daniels, Stephen. *Fields of Vision: Landscape Imagery and National Identity in England and the United States*. Cambridge: Polity Press, 1993.

Darnton, Robert. *The Literary Underground of the Old Regime*. Cambridge, Mass.: Harvard University Press, 1985.

———. *The Forbidden Best-Sellers of Pre-revolutionary France*. New York: W. W. Norton, 1996.

Daudet, Alphonse. "Contes choisis." In *Lectures alsaciennes: Géographie, histoire, biographies*, ed. Christian Pfister. Paris: Librairie Armand Colin, 1919.

de Certeau, Michel. *The Practice of Everyday Life*. Trans. Steven F. Rendall. Berkeley: University of California Press, 1984.

———. *Une politique de la langue: La révolution française et les patois*. Paris: Gallimard, 2002.

de Dainville, François. "L'Alsace comme la voyaient les cartes anciennes." *Saisons d'Alsace* 22 (1967): 153–76.

de Dietrich, Albert. *Alsaciens corrigeons notre accent!* Paris: Berger-Levrault, 1917.

Delahache, Georges. *Alsace-Lorraine: La carte au liseré vert*. Paris: Hachette, 1918.

de Morlet, Charles-Gabriel. *Topographie des Gaules: Notice sur les voies romaines du département du Bas-Rhin*. Strasbourg: Imprimerie de Veuve Berger-Levrault, 1861.

Denis, Marie-Noële. "Le dialecte alsacien: État des lieux." *Ethnologie française* 33, no. 3 (July 2003): 363–71.

Dipper, Christof, and Ute Schneider, eds. *Kartenwelten: Der Raum und seine Repräsentation in der Neuzeit*. Darmstadt: Primus, 2006.

Dollinger, Ferdinand. "A quelle race apartiennent les Alsaciens?" *Revue alsacienne illustrée* 5 (1903): 1–9.

Donan, Hastings, and Thomas M. Wilson, eds. *Border Identities: Nation and State at International Frontiers*. Cambridge: Cambridge University Press, 1998.

Dreyfus, Jean-Marc. "Eine Grenze in Ruinen: Zur Symbolik der Gipfel in den Vogesen." In *Wiedergewonnene Geschichte: Zur Aneignung von Vergangenheit in den Zwischenräumen Mitteleuropas*, ed. Peter Oliver Loew, Christian Pletzing, and Thomas Serrier. Wiesbaden: Harrassowitz Verlag, 2006.

Dufrenoy, Jules-Edouard. *Journal du siège de Strasbourg: Texte inédit présenté par Emmanuel Amougou*. Paris: L'Harmattan, 2004.

Duncan, James, and David Ley, eds. *Place/Culture/Representation*. London: Routledge, 1993.

Eckert, G. M. *Bilder aus dem Elsass*. Heidelberg: Verlag von Fr. Bassermann, 1874.

Eder, Susanne, and Martin Sandtner. "Common Spirit on the Upper Rhine Valley?" In *Boundaries and Place: European Borderlands in Geographical Context*, ed. David H. Kaplan and Jouni Häkli. New York: Rowman and Littlefield, 2002.

Edney, Matthew H. *Mapping an Empire: The Geographical Construction of British India, 1765–1843*. Chicago: University of Chicago Press, 1997.

Erckmann, E., and A. Chatrian. *L'ami Fritz*. Paris: Hachette, 1882.

Espagne, Michel. *Les transfers culturels franco-allemands*. Paris: Presses universitaires de France, 1999.

Espenhorst, Jürgen. *Petermann's Planet: A Guide to German Handatlases and Their Siblings throughout the World, 1800–1950*. Vol. 1. Trans. George R. Crossman. Schwerte: Pangaea Verlag, 2003.

Febvre, Lucien. *A Geographical Introduction to History*. Trans. E. G. Mountford and J. H. Paxton. New York: Barnes and Noble, 1966.

———. "*Frontière*: The Word and the Concept." In *A New Kind of History: From the Writings of Febvre*, ed. Peter Burke. London: Routledge, 1973.

———. Le Rhin: Histoire, mythes et réalités. Paris: Perrin, 1997.

Finsterwalder, Rüdiger. *Zur Entwicklung der bayerischen Kartographie von ihren Anfängen bis zum Beginn der amtlichen Landesaufnahme*. Munich: Verlag der Bayerischen Akademie der Wissenschaften, 1967.

Fischbach, Gustave. *Le siège et le bombardement de Strasbourg*. Strasbourg: Chez les principaux librairies, 1870.

Fischer, Albert. *Daniel Specklin aus Strassburg: Festungsbaumeister, Ingenieur und Kartograph*. Sigmarien: Jan Thorbecke Verlag, 1996.

Fischer, Christopher. *Alsace to the Alsatians? Visions and Divisions of Alsatian Regionalism, 1870–1939*. New York: Berghahn, 2010.

Ford, Caroline. *Creating the Nation in Provincial France: Religion and Political Identity in Brittany*. Princeton: Princeton University Press, 1993.

Foucault, Michel. *Discipline and Punish*. Trans. Alan Sheridan. New York: Vintage, 1995.

Fraipont, Gustave. *Les montagnes de France: Les Vosges*. Paris: Librairie Renouard, 1894.

Gallois, Lucien. "Un atlas d'Alsace-Lorraine." *Annales de géographie* 1 (September 1932): 518–23.

Gardt, Andreas, ed. *Sprachgeschichte als Kulturgeschichte*. Berlin: De Gruyter, 1999.

Geary, Patrick. *The Myth of Nations: The Medieval Origins of Europe*. Princeton: Princeton University Press, 2002.

Gerock, J. E. *La formation des départements du Haut-Rhin et du Bas-Rhin en 1789*. Thann: Imprimerie du Journal de Thann, 1925.

Gerson, Stéphane. *The Pride of Place: Local Memories and Political Culture in Nineteenth-Century France*. Ithaca: Cornell University Press, 2003.

Godlewska, Anne. *Geography Unbound: French Geographic Science from Cassini to Humboldt*. Chicago: University of Chicago Press, 1994.

Goethe, Johann Wolfgang von. *Aus meinem Leben, Dichtung und Wahrheit*. Berlin: H. Seemann Nachfilger, 1900.

Goodfellow, Samuel Huston. *Between the Swastika and the Cross of Lorraine: Fascisms in Interwar Alsace*. DeKalb: Northern Illinois University Press, 1999.

Gosewinkel, Dieter. *Figurationen des Staates in Deutschland und Frankreich, 1870–1945*. Munich: R. Oldenbourg, 2006.

Grenacher, Franz. "Current Knowledge of Alsatian Cartography." *Imago Mundi* 18 (1964): 60–77.

Grimm, Jacob, and Wilhelm Grimm. *Märchen der Brüder Grimm*. Weinheim: Beltz und Gelberg, 1995.

Guévremont, Elisabeth. "Les architectes travaillant à Strasbourg durant l'entre-deux-guerres, 1918–1939." Master's thesis, University of Strasbourg, 1997.

Gugerli, David, and Daniel Speich. *Topografien der Nation: Politik, kartographische Ordnung und Landschaft im 19. Jahrhundert*. Zurich: Chronos, 2002.

Guides Illustrés Michelin des Champs de Bataille. *Metz et la bataille de Morhange*. Clermont-Ferrand: Michelin & Co., 1919.

———. *L'Alsace et les combats des Vosges, 1914–1918*. Clermont-Ferrand: Michelin & Co., 1919.

Guides Michelin. *Strasbourg*. Clermont-Ferrand: Michelin, 1921.

Guiomar, Jean-Yves. "Vidal de la Blache's Geography of France." In *Realms of Memory: The Construction of the French Past*, ed. Pierre Nora, trans. Arthur Goldhammer, vol. 1. New York: Columbia University Press, 1997.

Gyss, Auguste. "L'historique de la Tour du Champ du Feu." *Les Vosges*, no. 4 (1984).

Habermas, Jürgen. *The Structural Transformation of the Public Sphere*. Trans. Thomas Burger. Cambridge, Mass.: MIT Press, 1989.

Habermeyer, Alphons. *Die topographische Landesaufnahme von Bayern im Wandel der Zeit*. Stuttgart: Verlag Konrad Wittwer, 1993.

Hahn, F. J., and J. L. Pinol. "La mobilité d'une grande ville: Strasbourg 1870 à 1940." *Annales de démographie historique* 1 (1995): 197–210.

Harley, J. B. *The New Nature of Maps: Essays in the History of Cartography*. Baltimore: Johns Hopkins University Press, 2001.

Harley, J. B., and David Woodward, eds. *The History of Cartography*. 6 vols. Chicago: University of Chicago Press, 1987–present.

Harp, Stephen. *Learning to Be Loyal: Primary Schooling as Nation Building in Alsace and Lorraine, 1850–1940*. DeKalb: Northern Illinois University Press, 1998.

———. *Marketing Michelin: Advertising and Cultural Identity in Twentieth-Century France*. Baltimore: Johns Hopkins University Press, 2001.

Harrison, Charles. "The Effects of Landscape." In *Landscape and Power*, ed. W. J. T. Mitchell. Chicago: University of Chicago Press, 2002.

Harvey, David. *Paris: Capital of Modernity*. New York: Routledge, 2005.

Harvey, David Allen. "Lost Children or Enemy Aliens? Classifying the Population of Alsace after the First World War." *Journal of Contemporary History* 34, no. 4 (October 1999): 537–54.

———. *Constructing Class and Nationality in Alsace, 1830–1945*. DeKalb: Northern Illinois University Press, 2001.

Haubrichs, Wolfgang, and Reinhard Schneider, eds. *Grenzen und Grenzregionen*. Saarbrücken: Saarbrücker Druckerei und Verlag, 1993.

Haupt, Heinz-Gerhard, and Jürgen Kocka. "Comparative History: Methods, Aims, Problems." In *Comparison and History: Europe in Cross-National Perspective*, ed. Maura O'Connor and Deborah Cohen. New York: Routledge, 2004.

Heffernan, Mark. "The Cartography of the Fourth Estate: Mapping the New Imperialism in British and French Newspapers, 1875–1925." In *The Imperial Map: Cartography and the Mastery of Empire*, ed. James R. Akerman. Chicago: University of Chicago Press, 2009.

Hennigé, Fr. E. *L'Alsace: Géographie locale et régionale à l'usage des écoles primaires des lycées et des collèges avec modèle d'une monographie géographique du lieu de domicile, carte en couleurs, plans et croquis*. Colmar: Société alsacienne d'édition Alsatia, 1920.

Herb, Guntram Henrik. *Under the Map of Germany: Nationalism and Propaganda, 1918–1945*. New York: Routledge, 1996.

Herringa, Wilbert, and John Nerbonne. "Dialect Areas and Dialect Continua." *Language Variation and Change* 13 (2001): 375–400.

Herscher, Andrew. "Urban Formations of Difference: Borders and Cities in Post-1989 Europe." *European Review* 13, no. 2 (2005): 251–60.

Heuss-Knapp, Elly. *Souvenirs d'une allemande de Strasbourg, 1881–1934*. Trans. Jean-Yves Mariotte. Strasbourg: Éditions Oberlin, 1996.

Hirsch, Francine. *Empire of Nations: Ethnographic Knowledge and the Making of the Soviet Union*. Ithaca: Cornell University Press, 2005.

Hirschon, Renée. *Crossing the Aegean: An Appraisal of the 1923 Compulsory Population Exchange between Greece and Turkey*. New York: Berghahn Books, 2002.

Hobsbawm, E. J. *Nations and Nationalism since 1780: Programme, Myth, Reality*. Cambridge: Cambridge University Press, 1992.

Hopkin, David. "Identity in a Divided Province: The Folklorists of Lorraine, 1860–1960." *French Historical Studies* 23, no. 4 (2000): 639–82.

Hostetler, Laura. *Qing Colonial Enterprise: Ethnography and Cartography in Early Modern China*. Chicago: University of Chicago Press, 2001.

Hugo, Victor. *Le Rhin: Lettres à un ami*. Vol. 2. Paris: Imprimerie Nationale, 1985.

Hyde, Charles Cheney. "Notes on Rivers as Boundaries." *American Journal of International Law* 6 (1912): 901–9.

Hyman, Paula E. *The Emancipation of the Jews of Alsace: Acculturation and Tradition in the Nineteenth Century*. New Haven: Yale University Press, 1992.

Igersheim, François. *L'Alsace des notables, 1870–1914*. Strasbourg: Budderflade, 1981.

———. *L'Alsace et ses historiens, 1680–1914: La fabrique des monuments*. Strasbourg: Presses universitaires de Strasbourg, 2006.

Ingold, Tim. *Being Alive: Essays on Movement, Knowledge and Description*. London: Routledge, 2011.

Iriye, Akira. "Transnational History." *Contemporary European History* 13, no. 2 (2004): 211–22.

Jacob, Christian. *The Sovereign Map: Theoretical Approaches in Cartography throughout History*. Trans. Tom Conley. Chicago: University of Chicago Press, 2006.

Jonas, Stéphane, et al., eds. *Strasbourg, capitale du Reichsland Alsace-Lorraine et sa nouvelle université, 1871–1918*. Strasbourg: Éditions Oberlin, 1994.

Judson, Pieter. *Guardians of the Nation: Activists on the Language Frontiers of Imperial Austria*. Cambridge, Mass.: Harvard University Press, 2006.

Kaelble, Hartmut. *Nachbarn am Rhein: Entfremdung und Annäherung der französischen und deutschen Gesellschaft seit 1880*. Munich: Beck, 1991.

Kaplan, David, and Jouni Haelki, eds. *Boundaries and Place: European Borderlands in Geographical Context*. Oxford: Rowman and Littlefield, 2002.

Kerdilès Weiler, Angéla. *Limites urbaines de Strasbourg: Évolution et mutation*. Strasbourg: Publications de la société savante d'Alsace, 2005.

Kiepert, Heinrich. *Spezialkarte der deutsch-französische Grenzlände mit Angabe der Sprachgrenze*. Berlin: Dietrich Reimer, 1867.

———. "Die Sprachgrenze in Elsass-Lothringen." *Zeitschrift der Gesellschaft für Erdkunde zu Berlin* 9 (1874): 307–16.

Kirschleger, Frédéric. *Flore d'Alsace et des contrées limitrophes*. Vol. 3. Strasbourg: Imprimerie d'Ed. Huder, 1862.

Kivelson, Valerie. *Cartographies of Tsardom: The Land and Its Meanings in Seventeenth-Century Russia*. Ithaca: Cornell University Press, 2006.

Kocka, Jürgen. "Comparison and Beyond." *History and Theory* 42 (February 2003): 39–44.

König, Eugen. *Führer durch Strassburg und die Vogesen*. Strassburg: G. Fischbach, 1896.

Königlich Preussische Landesaufnahme. *Die Arbeiten der Königlich Preussischen Landes-Aufnahme*. Berlin: Königlich Preussische Landesaufnahme, 1893.

———. *Musterblatt und Zeichenerklärung für die topographischen und kartographischen Arbeiten im Masstabe 1:25,000*. Berlin: Königlich Preussische Landesaufnahme, 1913.

Konvitz, Josef W. *Cartography in France, 1660–1848: Science, Engineering, and Statecraft.* Chicago: University of Chicago Press, 1987.

Koselleck, Reinhart. *Kritik und Krise.* Freiburg: K. Alber, 1959.

Koshar, Rudy. "What Ought to Be Seen: Tourists' Guidebooks and National Identities in Modern Germany and Europe." *Journal of Contemporary History* 33, no. 3 (July 1998): 323–40.

Kraemer, J. *Album von Strassburg.* Strassburg: R. Schultz, 1878.

Kramsch, Olivier, and Barbara Hoopers, eds. *Cross-Border Governance in the European Union.* London: Routledge, 2004.

Krauss, Georg. "150 Jahre preussische Messtischblätter." *Zeitschrift für Vermessungswesen* 4 (April 1969): 125–35.

Kühne, Thomas, ed. *Raum und Geschichte.* Leinfelden-Echterdingen: DRW-Verlag, 2001.

Kunz, Martin. *Das Bild in der Blindenschule.* Kiel: Druck von Schmidt und Klaunig, 1891.

Kwan, Mei-Po. "Feminist Visualization: Re-envisioning GIS as a Method in Feminist Geographic Research." *Annals of the Association of American Geographers* 92, no. 4 (December 2002): 645–661.

Labbé, Morgane. "Le projet d'une statistique des nationalités discuté dans les sessions du Congrès International de Statistique (1853–1876)." In *Démographie et politique,* ed. Hervé le Bras et al. Dijon: Éditions Universitaires de Dijon, 1997.

———. "Dénombrer les nationalités en Prusse au XIXe siècle: Entre pratique d'administration locale et connaissance statistique de la population." *Annales de démographie historique* 1 (2003): 39–61.

———. "La carte ethnographique de l'empire autrichien: La multinationalité dans l'ordre des choses." *Monde des cartes* 180 (2004): 71–83.

———. "Les frontières de la nation allemande dans l'espace de la carte, du tableau statistique et de la narration." In *L'espace de l'Allemagne au XIXe siècle: Frontières, centres et question nationale,* ed. Catherine Maurer. Strasbourg: Presses universitaires de Strasbourg, 2007.

Laboulais, Isabelle. *Lectures et pratiques de l'espace, l'itinéraire de Coquebert de Montbret (1755–1831), savant et grand commis d'État.* Paris: Honoré Champion, 1999.

———. "Modalités de construction: Un savoir cartographique et mobilisation des réseaux de correspondants; Le cas des ego-documents de Charles-Étienne Coquebert de Montbret (1755–1831)." In *Nouvelles approches des espaces et réseaux rationnels,* ed. Pierre-Yves Beaurepaire and Dominique Taurisson. Montpellier: Université Montpellier III, 2003.

———. "Reading a Vision of Space: The Geographical Map Collection of Charles-Étienne Coquebert de Montbret (1755–1831)." *Imago Mundi* 56, no. 1 (2004): 48–66.

Lagerlöf, Selma. *The Wonderful Adventure of Nils.* Trans. Velma Swanston Howard. New York: Doubleday, 1913.

Landes, Joan. *Visualizing the Nation: Gender, Representation, and Revolution in Eighteenth-Century France.* Ithaca: Cornell University Press, 2001.

Latour, Bruno. "Visualization and Cognition: Drawing Things Together." *Knowledge and Society: Studies in the Sociology of Culture Past and Present* 6 (1986): 1–40.

Lebovics, Herman. *True France: The Wars over Cultural Identity, 1900–1945.* Ithaca: Cornell University Press, 1992.

le Bras, Hervé, et al., eds. *Démographie et politique.* Dijon: Éditions universitaires de Dijon, 1997.

Lefebvre, Henri. *The Production of Space.* Trans. D. Nicholson-Smith. Oxford: Basil Blackwell, 1991.

Lefftz, Joseph. *Das Volkslied im Elsass.* Vol. 1. Colmar: Editions Alsatia, 1966.

Legoyt, Alfred. *La France et l'étranger: Études d'une statistique comparée.* Strasbourg: Berger-Levrault, 1865.

Lekan, Thomas. *Imagining the Nation in Nature: Landscape Preservation and German Identity, 1885–1945*. Cambridge, Mass.: Harvard University Press, 2004.

Levasseur, E., and Ch. Périgot. *Cartes pour server à l'intelligence de la France avec ses colonies*. Paris: Ch. Delagrave, 1874.

Levrault, L., Th. de Morville, and X. Mossmann. *Musée pittoresque et historique de l'Alsace: Dessins et illustrations par J. Rothmuller*. Colmar: J. Rothmuller, Editeur, 1863.

Lévy, Paul. *Histoire linguistique d'Alsace et de Lorraine*. 2 vols. Strasbourg: Imprimerie Alsacienne, 1929.

Liebich, L. "Esquisse d'une carte linguistique de l'Alsace." *Revue d'Alsace* 2, no. 2 (1861): 337–43.

Lienhart, H., and E. Martin. *Wörterbuch der elsässischen Mundarten*. Vol. 1. Strassburg: Verlag von Karl J. Trübner, 1899.

Lightman, Bernard. *Victorian Popularizers of Science: Designing Nature for New Audiences*. Chicago: University of Chicago Press, 2007.

Liulevicius, Vejas. *War Land on the Eastern Front: Culture, National Identity, and German Occupation in World War I*. Cambridge: Cambridge University Press, 2000.

Livet, Georges, and Francis Rapp, eds. *Histoire de Strasbourg des origines à nos jours*. 4 vols. Strasbourg: Éditions des Dernières nouvelles de Strasbourg, 1980–81.

———. *Histoire de Strasbourg*. Toulouse: Privat, 1987.

Lloyd, David. *Battlefield Tourism: Pilgrimage and Commemoration of the Great War in Britain, Australia and Canada, 1919–1939*. Oxford: Berg, 1998.

Loew, Peter Oliver, Christian Pletzing, and Thomas Serrier, eds. *Wiedergewonnene Geschichte: Zur Aneignung von Vergangenheit in den Zwischenräumen Mitteleuropas*. Wiesbaden: Harrassowitz Verlag, 2006.

Luthmer, Hans. "Der Vogesenclub von 1872–1918." In *Das Reichsland Elsass-Lothringen, 1871–1918*, vol. 3, ed. Georg Wolfram. Frankfurt: Selbstverlag des Elsass-Lothringen Instituts, 1934.

Maas, Annette. "Stadtplanung und Öffentlichkeit in Strassburg (1870–1918/25): Vom Nationalbewusstsein zur regionalen Identität städtischer Interessengruppen." In *Grenzstadt Strassburg: Stadtplanung, kommunale Wohnungspolitik und Öffentlichkeit, 1870–1940*, ed. Christof Cornelissen et al. St. Ingbert: Röhrig Universitätsverlag, 1997.

Marin, Louis. *On Representation*. Trans. Catherine Porter. Stanford: Stanford University Press, 2001.

Martin, Henri. *Le Vieil Armand, 1915*. Paris: Payot, 1936.

Massey, Doreen. *For Space*. London: Sage Press, 2005.

Maurer, Catherine, ed., *L'espace de l'Allemagne au XIXe siècle: Frontières, centres et question nationale*. Strasbourg: Presses universitaires de Strasbourg, 2007.

Mayeur, Jean-Marie. "A Frontier Memory: Alsace." In *Rethinking France: Les Lieux de Mémoire*, ed. Pierre Nora, trans. Mary Seidman Trouille, 3 vols., vol. 2. Chicago: University of Chicago Press, 2006.

McCoy, Rebecca. "The Culture of Accommodation: Religion, Language, and Politics in an Alsatian Community, 1648–1870." Ph.D. diss., University of North Carolina, Chapel Hill, 1992.

Mehmel, Astrid. "Deutsche Revisionspolitik in der Geographie nach dem Ersten Weltkrieg." *Geographische Rundschau* 9 (September 1995): 498–505.

Mentz, Ferdinand. "Die Ortsnamenverdeutschung in Elsass-Lothringen." *Zeitschrift des Allgemeinen Deutschen Sprachvereins* 31, no. 1 (January 1916): 4–8.

Merriman, John. *The Margins of City Life*. New York: Oxford University Press, 1991.

Michael, Bernardo. *Statemaking and Territory in South Asia: Lessons from the Anglo-Gorkha War (1814–1816)*. London: Anthem Press, 2012.

Michel, A. *Précis de géographie à l'usage des classes intermédiares de l'école primaire publique de Mulhouse*. Mulhouse: Imprimerie de J. P. Risler, 1869.

Ministère de la Défense Nationale et de la Guerre. *Le Service géographique de l'armée: Son histoire, son organisation, ses travaux.* Paris: Imprimerie du Service géographique de l'armée, 1938.

Ministère de la Guerre. *Instruction sur les cartes et plans directeurs de guerre (première partie).* Paris: Service géographique de l'armée, 1925.

Mitchell, W. J. T., ed. *Landscape and Power.* Chicago: University of Chicago Press, 1994.

——. "Introduction." In *Landscape and Power*, ed. W. J. T. Mitchell. Chicago: University of Chicago Press, 2002.

Monmonier, Mark. *How to Lie with Maps.* Chicago: University of Chicago Press, 1996.

Moser, G., and G. Kaufmann. *Geographische Faustzeichnung als Grundlage für einen methodischen Unterricht in der Geographie.* Strassburg: R. Schultz u. Comp., 1875.

Mukerji, Chandra. *Territorial Ambitions and the Gardens of Versailles.* Cambridge: Cambridge University Press, 1997.

Muller, Claude, and Bernard Vogler. *Catholiques et protestants en Alsace: Le simultaneum de 1802 à 1892.* Strasbourg: Librairie ISTRA, 1983.

Mündel, Curt. *Die Vogesen: Reisehandbuch für Elsass-Lothringen und angrenzende Gebiete.* Strassburg: Verlag von Karl J. Trübner, 1881.

Mundy, Barbara. *The Mapping of New Spain.* Chicago: University of Chicago Press, 1996.

Murphy, Alexander P. "Rethinking Multi-level Governance in a Changing European Union: Why Metageography and Territoriality Matter." *Geo-Journal* 72, no. 1/2 (2008): 7–18.

Naeher, J. *Panorama vom Donon im Elsass.* Strassburg im Elsass: Verlag von J. H. Ed. Heitz, 1888.

Nicolet, Claude. *Space, Geography, and Politics in the Early Roman Empire.* Trans. Hélène Leclerc. Ann Arbor: University of Michigan Press, 1991.

Nohlen, Klaus. *Construire une capitale: Strasbourg impérial de 1870 à 1918; Les bâtiments officiels de la Place Impériale.* Strasbourg: Publications de la société savante d'Alsace, 1997.

——. "Das Bild der Stadt Strassburg zur Reichslandszeit: Historischer Kern versus Neustadt." *Revue d'Alsace* 131 (2005): 139–63.

Nora, Pierre, ed. *Les lieux de mémoire.* 3 vols. Paris: Gallimard, 1984–92.

——. "Comment écrire l'histoire de France?" In *Les lieux de mémoire*, ed. Pierre Nora, vol. 3. Paris: Gallimard, 1992.

——, ed. *Realms of Memory: The Construction of the French Past.* Trans. Arthur Goldhammer. 3 vols. New York: Columbia University Press, 1996–98.

——, ed. *Rethinking France: Les Lieux de Mémoire.* Trans. Mary Seidman Trouille. 3 vols. Chicago: University of Chicago Press, 1999–2009.

Nordman, Daniel. *Frontières de France: De l'espace au territoire, XVI–XIXe siècle.* Paris: Gallimard, 1998.

O'Connor, Maura, and Deborah Cohen, eds. *Comparison and History: Europe in Cross-National Perspective.* New York: Routledge, 2004.

Oettermann, Stephan. *The Panorama: History of a Mass Medium.* New York: Zone Books, 1997.

Olson, Kory. "Creating Map Readers: Republican Geographic and Cartographic Discourse in G. Bruno's (Augustine Fouillée) 1905 Le Tour de la France par deux enfants." *Modern and Contemporary France* 19, no. 1 (February 2011): 37–51.

Ozouf, Jacques, and Mona Ozouf. "*Le tour de la France par deux enfants*: The Little Red Book of the Republic." In *Realms of Memory: The Construction of the French Past*, ed. Pierre Nora, trans. Arthur Goldhammer, vol. 2. New York: Columbia University Press, 1997.

Ozouf-Marignier, Marie-Vic. *La formation des départements: La représentation du territoire français à la fin du 18e siècle.* Paris: Éditions de l'École des hautes études en sciences sociales, 1989.

Palsky, Gilles. *Des chiffres et des cartes: La cartographie quantitative au XIXe siècle*. Paris: Comité des travaux historiques et scientifiques, 1996.

———. "Emmanuel de Martonne and the Ethnological Cartography of Central Europe (1917–1920)." *Imago Mundi* 54 (2002): 111–19.

Passman, Elana. "The Cultivation of Friendship: French and German Cultural Cooperation, 1925–1954." Ph.D. diss., University of North Carolina, Chapel Hill, 2009.

Pedley, Mary Sponberg. *The Commerce of Cartography: Making and Marketing Maps in Eighteenth-Century France and England*. Chicago: University of Chicago Press, 2005.

Pelletier, Monique. *Les cartes des Cassini: La science au service de l'État et des régions*. Paris: Éditions du C.T.H.S., 2002.

Pelletier, Monique, and Henriette Ozanne. *Portraits de la France: Les cartes, témoins de l'histoire*. Paris: Hachette, 1995.

Peluso, Nancy Lee. "Whose Woods Are These? Counter-mapping Forest Territories in Kalimantan, Indonesia." *Antipode* 27, no. 4 (October 1995): 383–406.

Penny, H. Glenn. *Objects of Culture: Ethnology and Ethnographic Museums in Imperial Germany*. Chapel Hill: University of North Carolina Press, 2002.

Perrot, Jean-Claude. *L'âge d'or de la statistique régionale française (an IV–1804)*. Paris: Société des études Robespierristes, 1977.

Petermann, Augustus. "Das General-Gouvernement Elsass und die deutsch-französische Sprachgrenze." *Pettermanns Geographische Mittheilungen* 16 (1870): plate 22.

Petite géographie de l'Alsace et la Lorraine à l'usage des écoles. Strasbourg: A. Viz et Co. Librairies-Editeurs, 1919.

Pétry, François. "Invention du paysage et identité aux XIXème et XXème siècles en Alsace." *Revue d'Alsace* 131 (2005): 277–364.

Petto, Christine Marie. *When France Was King of Cartography: The Patronage and Production of Maps in Early Modern France*. New York: Lexington Books, 2007.

Pfister, Christian. *Lectures alsaciennes: Géographie, histoire, biographies*. Paris: Librairie Armand Colin, 1919.

Piton, Frédéric. *Strasbourg illustré ou panorama pittoresque, historique et statistique de Strasbourg et de ses environs*. Strasbourg: Chez l'auteur, rue du Temple Neuf, 15, et chez les principaux libraires, 1855.

———. *Promenades en Alsace: Monographies historiques, archéologiques et statistiques; Ribeauvillé et ses environs*. Strasbourg: Chez l'auteur, 1856.

———. *Siège de Strasbourg: Journal d'un assiégé*. Paris: Charles Schlaeber, 1900.

Pomian, Krzysztof. "Francs et Gaulois." In *Les lieux de mémoire*, ed. Pierre Nora, vol. 3. Paris: Gallimard, 1992.

Pounds, Norman. "France and 'Les Limites Naturelles' from the Seventeenth to the Twentieth Centuries." *Annals of the Association of American Geographers* 44, no. 1 (March 1954): 51–62.

Pratt, Mary Louise. *Imperial Eyes: Travel Writing and Transculturation*. New York: Routledge, 1992.

Purdue, Peter. "Boundaries, Maps, and Movement: The Chinese, Russian, and Mongolian Empires in Early Modern Eurasia." *International History Review* 20, no. 2 (June 1998): 263–86.

Ramaswamy, Sumathi. *The Goddess and the Nation: Mapping Mother India*. Durham: Duke University Press, 2010.

Raphael, Samuel. *Theatres of Memory*. Vol. 1. London: Verso, 1996.

Ratzel, Friedrich. *Politische Geographie*. Munich: Verlag von R. Oldenbourg, 1897.

Rebert, Paula. *La Gran Linea: Mapping the United States-Mexico Boundary, 1849–1857*. Austin: University of Texas Press, 2001.

Renan, Ernest. *"Qu-est-ce qu'une nation?" Conférence faite en Sorbonne, le 11 mars 1882*. Paris: Calmann Lévy, 1882.

Reuss, Rodolphe. *Vieux noms et rues nouvelles de Strasbourg: Causeries biographiques d'un flâ- neur.* Strasbourg: Treuttel et Wuertz, Editeurs, 1883.

Revel, Jacques, ed. *L'espace français.* Paris: Editions du Seuil, 1989.

Robb, Graham. *The Discovery of France: A Historical Geography from the Revolution to the First World War.* New York: W. W. Norton, 2007.

Robinson, Richard. *Narratives of the European Border: A History of Nowhere.* New York: Palgrave MacMillan, 2007.

Roos, Karl. *Die Fremdwörter in den elsässischen Mundarten: Ein Beitrage zur elsässischen Dia- lektforschung.* Strassburg: Universitäts-Buchdrückerei von J. H. Ed. Heitz, 1903.

Roth, François. "La frontière franco-allemande 1871–1918." In *Grenzen und Grenzregionen,* ed. Wolfgang Haubrichs and Reinhard Schneider. Saarbrücken: Saarbrücker Druck und Verlag, 1993.

Rousseau, Jean-Jacques. *Émile.* Paris: Hachette: 1880.

Rudolph, E. *Landeskunde des Reichslandes Elsass-Lothringen.* Breslau: Königliche Universitäts- und Verlagsbuchhandlung, 1907.

Sack, Robert David. *Human Territoriality: Its Theory and History.* Cambridge: Cambridge University Press, 1986.

Sahlins, Peter. *Boundaries: The Making of France and Spain in the Pyrenees.* Berkeley: Univer- sity of California Press, 1989.

———. "Natural Frontiers Revisited: France's Boundaries since the Seventeenth Century." *American Historical Review* 95, no. 5 (December 1990): 1423–51.

Said, Edward W. *Orientalism.* New York: Vintage, 1979.

Saint-Gérard, Jacques-Philippe, ed. *Mutations et sclérose: La langue française, 1789–1840.* Stuttgart: F. Steiner, 1993.

Sanders, Ruth. *German: Biography of a Language.* Oxford: Oxford University Press, 2010.

Schama, Simon. *Landscape and Memory.* New York: Knopf, 1995.

Scharf, Th. *Geographie für die Volksschulen von Elsass-Lothringen.* Metz: Schulbuchhandlung von Wwe. Alcan, 1876.

Schlögel, Karl. *Im Raume Lesen Wir die Zeit.* Munich: Carl Hanser Verlag, 2003.

Schoebel, Charles. *La question d'Alsace au point de vue éthnographique.* Paris: Sandoz et Fisch- bacher, Éditeurs, 1872.

Schrader, F. *Atlas de géographie moderne.* Paris: Hachette, 1890.

Schrader, F., and L. Gallouédec. *Géographie de la France et de ses colonies.* Paris: Hachette, 1899.

Schricker, August. *Strassburg und Umgebung.* Zürich: J. Laurencie, 1890.

Schroeder-Hohenwarth, J. "Die preussische Landesaufnahme von 1816–1875." *Nachrichten aus dem Karten- und Vermessungswesen* 5 (1958): 5–59.

Schulten, Susan. *The Geographical Imagination in America, 1880–1950.* Chicago: University of Chicago Press, 2001.

———. *Mapping the Nation: History and Cartography in Nineteenth-Century America.* Chicago: University of Chicago Press, 2012.

Schwartz, Vanessa, and Jeannene Przyblyski, eds. *The Nineteenth-Century Visual Culture Reader.* New York: Routledge, 2004.

Scott, James C. *Seeing like a State: How Certain Schemes to Improve the Human Condition Have Failed.* New Haven: Yale University Press, 1998.

Seegel, Steven. *Mapping Europe's Borderlands: Russian Cartography in the Age of Empire.* Chi- cago: University of Chicago Press, 2012.

Serrier, Thomas. *Entre Allemagne et Pologne: Nations et identités frontalières, 1848–1914.* Paris: Belin, 2002.

Service géographique de l'armée. *Rapports sur les travaux exécutés.* Paris: Librairie militaire de L. Baucoin, 1910.

Seyboth, Adolphe. *Das Alte Strassburg vom 13. Jahrhundert bis zum Jahre 1870: Geschichtliche Topographie nach den Urkunden und Chroniken.* Strassburg: J. H. Heitz, 1890.

———. *Ansichten des Alten Strassburg: Fünfzig Tafeln mit erklärendem Text.* Strassburg: J. H. Heitz, 1891.

Shepard, Todd. *The Invention of Decolonization: The Algerian War and the Remaking of France.* Ithaca: Cornell University Press, 2006.

Sherman, Daniel. *The Construction of Memory in Interwar France.* Chicago: University of Chicago Press, 2001.

Siegel, Mona. *The Moral Disarmament of France: Education, Pacifism, and Patriotism, 1914–1940.* Cambridge: Cambridge University Press, 2004.

Silverman, Dan P. *Reluctant Union: Alsace-Lorraine and Imperial Germany, 1871–1918.* University Park: Pennsylvania State University Press, 1972.

Singer, Brian. "Cultural versus Contractual Nations: Rethinking Their Opposition." *History and Theory* 35, no. 3 (1996): 309–37.

Smith, Nathaniel B. "The Idea of the French Hexagon." *French Historical Studies* 6, no. 2 (Autumn 1969): 139–55.

Société de Géographie. *Congrès international des sciences géographiques tenu à Paris du 1er au 11 août 1875: Compte rendu des séances.* Paris: Imprimerie de E. Martinet, 1875.

Soja, Edward. *Postmodern Geographies: The Reassertion of Space in Critical Social Theory.* London: Verso, 1989.

Städtischen Statistischen Amt. *Verwaltungsbericht der Stadt Strassburg für die Zeit von 1870 bis 1888/89.* Strassburg: Elsässische Druckerei und Verlagsanstalt vorm. G. Fischbach, 1895.

Steinhoff, Anthony. *The Gods of the City: Protestantism and Religious Culture in Strasbourg, 1870–1914.* Boston: Brill, 2008.

St. John, Rachel. *Line in the Sand: A History of the Western U.S.-Mexico Border.* Princeton: Princeton University Press, 2012.

Struck, Bernhard. "Farben, Sprachen, Territorien: Die deutsch-polnische Grenzregion auf Karten des 19. Jahrhunderts." In *Kartenwelten: Der Raum und seine Repräsentation in der Neuzeit*, ed. Christof Dipper and Ute Schneider. Darmstadt: Primus, 2006.

Taverne, Ed, and Cor Wagenaar. "Border Cities: Contested Identities of the European City." *European Review* 13, no. 2 (2005): 201–6.

Taylor, Isadore Baron, and Charles Nodier. *Voyages pittoresques et romantiques dans l'ancienne France.* Paris: Didot, 1833.

Thiesse, Anne-Marie. *Ils apprenaient la France: L'exaltation des régions dans le discours patriotique.* Paris: Éditions de la Maison des Sciences de l'Homme, 1997.

This, Constant. *Die deutsch-französische Sprachgrenze in Lothringen.* Strassburg: J. H. Ed. Heitz, 1887.

———. *Die deutsch-französische Sprachgrenze im Elsass.* Strassburg: J. H. Ed. Heitz, 1888.

Thompson, E. P. *Customs in Common.* New York: New Press, 1991.

Towson, Michael. *Mother-Tongue and Fatherland: Language and Politics in Germany.* New York: St. Martin's Press, 1992.

Truett, Samuel. *Fugitive Landscapes: The Forgotten History of the U.S.-Mexico Borderland.* New Haven: Yale University Press, 2008.

Tuan, Yi-Fu. *Space and Place: The Perspective of Experience.* Minneapolis: University of Minnesota Press, 2001.

Verlagskatalog der J. Boltzeschen Buchhandlung O. H. Gebweiler, 1870–1910. Leipzig: Buch- u. Kunstdruckerei Breitkopf & Haertel, 1910.

Vidal de la Blache, Paul. *Tableau de la géographie de France.* Paris: Hachette, 1903.

Vigato, Jean-Claude. "L'architecture régionaliste de 1900 à 1930." *Revue d'Alsace* 131 (2005): 165–88.

Vlossak, Elizabeth. *Marianne or Germania? Nationalizing Women in Alsace, 1870–1946.* Oxford: Oxford University Press, 2010.

Vogler, Bernard. *Histoire culturelle de l'Alsace: Du Moyen Age à nos jours, les très riches heures d'une région frontière*. Strasbourg: La Nuée bleue, 1994.

von Morozowicz, O. *Königlich Preussische Landes-Aufnahme*. Berlin: Königlich Preussische Landes-Aufnahme, 1878.

von Thudichum, Friedrich. *Historisch-statistiche Grundkarten: Denkschrift*. Tübingen: Verlag der H. Laupp'schen Buchhandlung, 1892.

Wahl, Alfred, and Jean-Claude Richez. *L'Alsace entre France et Allemagne, 1850–1950*. Paris: Hachette, 1994.

Walser Smith, Helmut. *The Continuities of German History: Nation, Religion, and Race across the Long Nineteenth Century*. Cambridge: Cambridge University Press, 2008.

Weber, Eugen. *Peasants into Frenchmen: The Modernization of Rural France, 1870–1914*. Stanford: Stanford University Press, 1976.

———. *My France: Politics, Culture, Myth*. Cambridge, Mass.: Harvard University Press, 1991.

Wehler, Hans-Ulrich. *Krisenherde des Kaiserreichs 1871–1918*. Göttingen: Vandenhoeck und Ruprecht, 1970.

Weick, Georg. *Heimatkunde (Heimat und Kreis): Ein Hilfsbüchlein für den ersten Geographieunterricht*. Zabern: Druck und Verlag der Schulbuchhandlung H. Fuchs, 1894.

Werner, Michel, and Bénédicte Zimmermann. "Beyond Comparison: *Histoire croisée* and the Challenge of Reflexivity." *History and Theory* 45, no. 1 (February 2006): 30–50.

Western, John. *Cosmopolitan Europe: A Strasbourg Self-Portrait*. London: Ashgate, 2012.

Whalen, Philip, and Patrick Young, eds. *Place and Locality in Modern France*. New York: Bloomsbury Press, 2014.

Wickerscheimer, G. *Geographie von Elsass-Lothringen bearbeitet von einem Mitgliede der Strassburger Lehrer-Conferenz*. Strassburg: Druck und Verlag von J. H. Eduard Heitz, 1872.

Wigen, Kären. *A Malleable Map: Geographies of Restoration in Central Japan, 1600–1912*. Berkeley: University of California Press, 2010.

Williams, John Alexander. *Turning to Nature in Germany: Hiking, Nudism, and Conservation*. Stanford: Stanford University Press, 2007.

Winichakul, Thongchai. *Siam Mapped: A History of the Geo-body of a Nation*. Honolulu: University of Hawaii Press, 1994.

Wolfram, Georg, ed. *Das Reichsland Elsass-Lothringen, 1871–1918*. Vol. 3. Frankfurt: Selbstverlag des Elsass-Lothringischen Instituts, 1934.

Wolfram, Georg, and Werner Gley, eds. *Elsass-Lothringer Atlas: Landes-kunde, Geschichte, Kultur und Wirtschaft Elsass-Lothrigens*. Frankfurt: Selbstverlag des Elsass-Lothringischen Instituts, 1931.

Woodward, David, ed. *Art and Cartography: Six Historical Essays*. Chicago: University of Chicago Press, 1987.

Yildirim, Onur. *Diplomacy and Displacement: Reconstructing the Turco-Greek Exchange of Populations, 1922–1934*. New York: Routledge: 2006.

Yonemoto, Marcia. *Mapping Early Modern Japan: Space, Place, and Culture in the Tokugawa Period (1603–1868)*. Berkeley: University of California Press, 2003.

Young, Patrick. *Enacting Brittany: Tourism and Culture in Provincial France, 1871–1939*. Surrey: Ashgate, 2012.

Zahra, Tara. "The Minority Problem: National Classification in the French and Czechoslovak Borderlands." *Contemporary European History* 17 (May 2008): 137–65.

Page numbers in boldface refer to illustrations.

de Martonne, Emmanuel, 66
de Morville, Th., 137–39, **138**
Dollinger, Ferdinand, 62
Donon Peak, 147–50, **148**
Doré, Gustave, 168–69
drawings, and territorial identity, 5, 107

education, and territorial identity. *See*
    schools, and territorial identity
Elsasskarte (Specklin), 21–24, **22**
Elsass-Lothringer Atlas (Gley and Wolfram),
    67–68
Elsass-Lothringische Jahrbuch, 67–68
Emerich, Dr., 181–82
Émile (Rousseau), 102
encyclopedias, 53–54
Enlightenment, and mapmaking, 5, 42,
    49–50
Erinnerungsblatt (map), 36, **37**
Esplanade (Strasbourg), 175, 176
ESPON (European Spatial Planning Obser-
    vation Network), 193
ethnic and racial borders, mapping of, 6,
    54–57, **56,** 64, 66–68, **plate 5**
ethnographers, and territorial identity, 5,
    48–49, 111–12
Ethnographic Map of the Austrian Monarchy
    (von Czoernig), 78
Europe: contemporary attitudes toward
    borders in, 192–95; ethnographic
    borders in, 54–59, 66–68, 87. *See also*
    specific European locations by name
European Court of Human Rights, 193
European Cyclists' Federation, 194
European Parliament, 193
European Quarter (Strasbourg), 193
European Union, 194
Euro-regions, 193–95
Euting, Julius, 141–42, 143, 149, 151
exhibitions, maps in, 9, 113–14, 192

fairy tales and folklore, 53, 75
Favre, Jules, 31–32
Febvre, Lucien, 68, 188
Fichte, Johann Gottfried, 75
Fischbach, Gustave, 169
Flanders, 10
Flemish language, 73
Foch, Ferdinand, 183
Ford, Caroline, 115
"Foreign Words in the Alsatian Dialects"
    (Roos), 88
Fouillée, Augustine, 100, 116–20, **117,** 121,
    125
Fraipont, Gustave, 149–50
France: capital and borderlands in, 10–12,
    20–21, 24–27, 93–101, 114–29, 163–73,

183–89; civilian debates about borders,
    49–52, 57–62, **61,** 188, **plates 7–8;**
    language borders in, 70–75, **plate 9;**
    state maps of, 19–21, 24–28, **25,** 41–45,
    44, plate 4; transfers of Alsace and
    Lorraine with Germany, 2–3, 11–12,
    24, 31–34, 60–61, 66–67, 80, 86
Franconian dialect, 70, 215n1
Franco-Prussian War, 11, 31, 59, 79, 114,
    115, **165,** 167–73, 209n62, 231n14
Frankfurt, Treaty of, 31–32, 34, 79, 132
Frankfurt Parliament of 1848, 75, 77
Fréland, 122–23, **123, 124, 125**
French Ethnographic Society, 58
French Geographical Society, 212n19
French-German Border Commission, 6, 32–
    34, **33,** 48
French language: dialects of, 70–71, 74, 87,
    215n1; as spoken in Alsace and Lor-
    raine, 55, 57–58, 70–89
French Revolution, 5, 11, 26, 28, 51, 72–73,
    115–16, 170, 223n40
French Statistics Administration, 57–58
Friedrich, Caspar David, 135–36, **135,** 149
Fröschweiler, 98, **98**

Gadolle, Pierre, 51–52
Gallois, Lucien, 66
Gaul, 50, 62, 131, 147–49
Generalstabskarte (General Staff map), 19–
    20, 36, 43
geo-body, concept of, 9–10, 48
geographers, and territorial identity, 5,
    48–49, 53–57, 59, 60–62, 66, 79–80, 87,
    101–14, 116–25, 188, 194
Geographical Institute (Weimar), 79
Geographical Service of the French Army,
    43–45
geographical societies, and territorial iden-
    tity, 53, 55, 81, 212n19
Geographical Society of Berlin, 53, 212n19
geography textbooks, 9, 60, 101–14, **108,**
    116–25, **117, 123, 124, 125, 126,** 194,
    223n48
German Architectural Association, 178–79
German Association of Teachers for the
    Blind, 107–8
German Borders against France, The (map),
    59, **plate 6**
German Forest Administration, 141
German language: dialects and forms of,
    70–71, 77, 215n1; as spoken in Alsace
    and Lorraine, 55, 57–58, 70–89
German Rhine, The (lithograph), 168–69
German School Association for the Preser-
    vation of Germans in Foreign Coun-
    tries, 79

statisticians, and territorial identity, 48–49, 57–59, 73–75, 77–80, 87
Statistics, Bureau of (France), 73–74, 75
Steinbach, 42
Stieler, Adolf, 77
Stieve, Richard, 140
Strasbourg: contemporary, 193; languages in, 72, 80, 184; in maps and images, 26, 113, 159–89, **162, 164, 165, 167, 168, 169, 171, 172, 173, 177, 178, 179, 182, 184, 186, plates 15–16;** name of, 80, 161, 174; siege of, 160, 167–73, **169, 171, 172, 173**, 209n62
Strasbourg, University of, 64, 67–68, 87, 123, 180, 188, 198
Strasbourg Beautification Society, 176
Strasbourg Tourist Commission, 126, 129
Striedbeck, Johannes, 161–63, **162**
St. Ulrich, Castle of, 110, 138, **138**
surveying, for mapmaking, 5, 8, 19–46, 96, 161
Switzerland, 83, 106, 187, 193, 194

Tabouis, Georges, 155
teachers, and territorial identity. *See* schools, and territorial identity
Temple-Neuf (Strasbourg), 170
territory, concepts of: cityscape, 159–89; geo-body, 9–10, 48; *grande patrie* (big homeland, nation-state), 6, 93–95, 114–29; *Heimat* (home), 6, 93–95, 101–14, 121–22, 141, 152, 182–83, 220n6; landscape, 131–58; *petite patrie* (little homeland, region), 6, 93–95, 114–29. *See also* borders and borderlands
Thiesse, Anne-Marie, 115
Thirty Years War, 23
This, Constant, 81–83, **82**
Topographical Map of the Rhine and Its Two Banks, 29–31, **30**
Topography of the Gauls project, 62, **63**
toponyms, and territorial identity, 7, 23, 37–38, 43, 73, 79–80, 86, 184, 185
Tour de France, idea of, 116–17
Touring-Club Vosgien, 152–53, 154
tourism, regional, 125, 131–58, 159
tourist guidebooks, 9, 42, 45, 137–46, 166, 176, 178, 180–81, 184–85, 226n2
Town Hall (Strasbourg), 181
Tranchot, Jean-Joseph, 28–29
transnational history, 2–3, 11–12, 24, 31–34, 60–61, 66–67, 80, 86, 221n7
transportation networks, mapping of, 38, 43, 105, 107, 110
travel narratives, use in mapmaking, 71, 74
triangulation, in mapmaking, 13, 24–28, **25,** 29, 38–39, **39,** 43, 146

TriRhena, 193
Tulla Rectification Projects, 187, 226n89
Turkey, 66
Tyrol, the, 12, 71, 79

Uhrich, Jean-Jacques, 209n62
universities, and mapmaking, 48–49, 53, 64, 66, 67–68
Upper Alsace. *See* Alsace and Lorraine

Vauban, Sébastien Le Prestre de, 156, 175
Versailles, Gardens of, 133–34, 224n65
Versailles, Treaty of, 66, 70, 87, 185, 187, 188
Vidal de la Blache, Paul, 66, 115, 120
Vienna Statistical Congress (1857), 57
Views of Old Strasbourg (Seyboth), 181, **182**
villages, in maps and images, 93–130. *See also specific villages by name*
Villot, Jean-Nicolas, **165**
Vogesenclub (Vosges Club), 132, 140–51, **143, 144, 145, 150,** 152, 153–54
"Vogesenlied" (song; Schmitt), 151
von Czoernig, Karl, 78
Von Decker Garrison (Strasbourg), 175
von Moeller, Eduard, 84
von Müffling, Friedrich Karl Ferdinand, 29
von Thudichum, Friedrich, 64–65
Vosges Mountains: as border, 3, 32–34, **33, 35,** 41–42, 59, 87; hiking trails in, 140–58; in maps and local images, 21, **22,** 110, 131–58, **135, 138, 142, 143, 144, 145, 148, 150, 156, 157, plate 14**
Vosges-Rhine Society (Société vogéso-rhénane), 139–40

Wanderer above a Sea of Fog (Friedrich), 135–36, **135,** 149
Wattwiller, 228n24
Weber, Eugen, 114–15
Weick, Georg, 103–5
Wigen, Kären, 11, 220n3
Wilhelm I (Germany), 2–3, 36, **37,** 170, 175, 184
Wilhelm II (Germany), 85–86, 151
Winichakul, Thongchai, 9–10
Winter, Charles, 172–73, **172, 173,** 181
Wissenschaftliche Institut der Elsass-Lothringer im Reich, 67–68
Wolfram, Georg, 67–68
World War I, 3, 40–42, 44–45, 65–68, 85–87, 153, 155–59, **156, 157,** 174, 183
World War II, 3, 68, 87–88, 129, 188–89, 192, 193
Württemberg, 10, 29, 35, 36, 209n50, 220n6

Zuber, Pierre, 154–55